ALSO BY JARED COHEN

Children of Jihad: A Young American's Travels
Among the Youth of the Middle East

One Hundred Days of Silence: America
and the Rwanda Genocide

The New Digital Age

The New Digital Age

RESHAPING the FUTURE of PEOPLE, NATIONS and BUSINESS

Eric Schmidt and **Jared Cohen**

ALFRED A. KNOPF   NEW YORK   2013

THIS IS A BORZOI BOOK
PUBLISHED BY ALFRED A. KNOPF

Copyright © 2013 by Google Inc. and Jared Cohen

All rights reserved. Published in the United States by Alfred A. Knopf,
a division of Random House, Inc., New York, and in Canada
by Random House of Canada Limited, Toronto.

www.aaknopf.com

Knopf, Borzoi Books, and the colophon are registered trademarks
of Random House, Inc.

Library of Congress Cataloging-in-Publication Data is on file at the Library of Congress
ISBN 978-0-307-95713-9

Jacket design by Peter Mendelsund

Manufactured in the United States of America
First Edition

For Rebecca, to whom we are grateful for her ideas

and support, and Aiden, of whom we are envious for

the technology he will get to see

We should all be concerned about the future because

we will have to spend the rest of our lives there.

—CHARLES F. KETTERING,
American inventor and businessman

CONTENTS

The New Digital Age

Introduction

The Internet is among the few things humans have built that they don't truly understand. What began as a means of electronic information transmission—room-sized computer to room-sized computer—has transformed into an omnipresent and endlessly multifaceted outlet for human energy and expression. It is at once intangible and in a constant state of mutation, growing larger and more complex with each passing second. It is a source for tremendous good and potentially dreadful evil, and we're only just beginning to witness its impact on the world stage.

The Internet is the largest experiment involving anarchy in history. Hundreds of millions of people are, each minute, creating and consuming an untold amount of digital content in an online world that is not truly bound by terrestrial laws. This new capacity for free expression and free movement of information has generated the rich virtual landscape we know today. Think of all the websites you've ever visited, all the e-mails you've sent and stories you've read online, all the facts you've learned and fictions you've encountered and debunked. Think of every relationship forged, every journey planned, every job found and every dream born, nurtured and implemented through this platform. Consider too what the lack of top-down control allows: the online scams, the bullying campaigns, the hate-group websites and the terrorist chat rooms. This is the Internet, the world's largest ungoverned space.

As this space grows larger, our understanding of nearly every aspect of life will change, from the minutiae of our daily lives to more fundamental questions about identity, relationships and even our own security. Through the power of technology, age-old obstacles to human interaction, like geography, language and limited information, are falling and a new wave of human creativity and potential is rising. Mass adoption of the Internet is driving one of the most exciting social, cultural and political transformations in history, and unlike earlier periods of change, this time the effects are fully global. Never before in history have so many people, from so many places, had so much power at their fingertips. And while this is hardly the first technology revolution in our history, it is the first that will make it possible for almost everybody to own, develop and disseminate real-time content without having to rely on intermediaries.

And we've barely left the starting blocks.

The proliferation of communication technologies has advanced at an unprecedented speed. In the first decade of the twenty-first century the number of people connected to the Internet worldwide increased from 350 million to more than 2 billion. In the same period, the number of mobile-phone subscribers rose from 750 million to well over 5 billion (it is now over 6 billion). Adoption of these technologies is spreading to the farthest reaches of the planet, and, in some parts of the world, at an accelerating rate.

By 2025, the majority of the world's population will, in one generation, have gone from having virtually no access to unfiltered information to accessing all of the world's information through a device that fits in the palm of the hand. If the current pace of technological innovation is maintained, most of the projected eight billion people on Earth will be online.

At every level of society, connectivity will continue to become more affordable and practical in substantial ways. People will have access to ubiquitous wireless Internet networks that are many times cheaper than they are now. We'll be more efficient, more productive and more creative. In the developing world, public wireless hot spots and high-speed home networks will reinforce each other, extending the online experience to places where people today don't even have land-line phones. Societies will leapfrog an entire generation of technology.

Eventually, the accoutrements of technologies we marvel at today will be sold in flea markets as antiques, like rotary phones before them.

And as adoption of these tools increases, so too will their speed and computing power. Moore's Law, the rule of thumb in the technology industry, tells us that processor chips—the small circuit boards that form the backbone of every computing device—double in speed every eighteen months. That means a computer in 2025 will be sixty-four times faster than it is in 2013. Another predictive law, this one of photonics (regarding the transmission of information), tells us that the amount of data coming out of fiber-optic cables, the fastest form of connectivity, doubles roughly every nine months. Even if these laws have natural limits, the promise of exponential growth unleashes possibilities in graphics and virtual reality that will make the online experience as real as real life, or perhaps even better. Imagine having the holodeck from the world of *Star Trek,* which was a fully immersive virtual-reality environment for those aboard a ship, but this one is able to both project a beach landscape and re-create a famous Elvis Presley performance in front of your eyes. Indeed, the next moments in our technological evolution promise to turn a host of popular science-fiction concepts into science facts: driverless cars, thought-controlled robotic motion, artificial intelligence (AI) and fully integrated augmented reality, which promises a visual overlay of digital information onto our physical environment. Such developments will join with and enhance elements of our natural world.

This is our future, and these remarkable things are already beginning to take shape. That is what makes working in the technology industry so exciting today. It's not just because we have a chance to invent and build amazing new devices or because of the scale of technological and intellectual challenges we will try to conquer; it's because of what these developments will mean for the world.

Communication technologies represent opportunities for cultural breakthroughs as well as technical ones. How we interact with others and how we view ourselves will continue to be influenced and driven by the online world around us. Our propensity for selective memory allows us to adopt new habits quickly and forget the ways we did things before. These days, it's hard to imagine a life without mobile devices. In a time of ubiquitous smart phones, you have insurance against forgetfulness, you have access to an entire world of ideas (even though some governments make it difficult), and you always have something to occupy your attention, although finding a way to do so usefully may still prove difficult and in some cases harder. The *smart* phone is aptly named.

As global connectivity continues its unprecedented advance, many old institutions and hierarchies will have to adapt or risk becoming obsolete, irrelevant to modern society. The struggles we see today in many businesses, large and small, are examples of the dramatic shift for society that lies ahead. Communication technologies will continue to change our institutions from within and without. We will increasingly reach, and relate to, people far beyond our own borders and language groups, sharing ideas, doing business and building genuine relationships.

The vast majority of us will increasingly find ourselves living, working and being governed in two worlds at once. In the virtual world we will all experience some kind of connectivity, quickly and through a variety of means and devices. In the physical world we will still have to contend with geography, randomness of birth (some born as rich people in rich countries, the majority as poor people in poor countries), bad luck and the good and bad sides of human nature. In this book we aim to demonstrate ways in which the virtual world can make the physical world better, worse or just different. Sometimes these worlds will constrain each other; sometimes they will clash; sometimes they will intensify, accelerate and exacerbate phenomena in the other world so that a difference in degree will become a difference in kind.

On the world stage, the most significant impact of the spread of communication technologies will be the way they help reallocate the concentration of power away from states and institutions and transfer it to individuals. Throughout history, the advent of new information technologies has often empowered successive waves of people at the expense of traditional power brokers, whether that meant the king, the church or the elites. Then as now, access to information and to new communication channels meant new opportunities to participate, to hold power to account and to direct the course of one's life with greater agency.

The spread of connectivity, particularly through Internet-enabled mobile phones, is certainly the most common and perhaps the most profound example of this shift in power, if only because of the scale.

Digital empowerment will be, for some, the first experience of empowerment in their lives, enabling them to be heard, counted and taken seriously—all because of an inexpensive device they can carry in their pocket. As a result, authoritarian governments will find their newly connected populations more difficult to control, repress and influence, while democratic states will be forced to include many more voices (individuals, organizations and companies) in their affairs. To be sure, governments will always find ways to use new levels of connectivity to their advantage, but because of the way current network technology is structured, it truly favors the citizens, in ways we will explore later.

So, will this transfer of power to individuals ultimately result in a safer world, or a more dangerous one? We can only wait and see. We have only begun to encounter the realities of a connected world: the good, the bad and the worrisome. The two of us have explored this question from different vantage points—one as a computer scientist and business executive and the other as a foreign-policy and national security expert—and we both know that the answer is not predetermined. The future will be shaped by how states, citizens, companies and institutions handle their new responsibilities.

In the past, international-relations theorists have debated the ambitions of states—some arguing that states maintain domestic and foreign policies that aim to maximize their power and security, while others suggest that additional factors, such as trade and information exchange, also affect state behavior. States' ambitions won't change, but their notions of how to achieve them will. They will have to practice two versions of their domestic and foreign policies—one for the physical, "real" world, and one for the virtual world that exists online. These policies may appear contradictory at times—governments might crack down in one realm while allowing certain behavior in another; they may go to war in cyberspace but maintain the peace in the physical world—but for states, they will represent attempts to deal with the new threats and challenges to their authority that connectivity enables.

For citizens, coming online means coming into possession of multiple identities in the physical *and* virtual worlds. In many ways, their virtual identities will come to supersede all others, as the trails they leave remain engraved online in perpetuity. And because what we post, e-mail, text and share online shapes the virtual identities of others, new forms of collective responsibility will have to come into effect.

For organizations and companies, opportunities and challenges will come hand in hand with global connectivity. A new level of accountability, driven by the people, will force these actors to rethink their existing operations and adapt their plans for the future, changing how they do things as well as how they present their activities to the public. They'll also find new competitors, as widespread technological inclusion levels the playing field for information, and therefore opportunity.

In the future, no person, from the most powerful to the weakest, will be insulated from what in many cases will be historic changes.

We two first met in the fall of 2009, under circumstances that made it easy to form a bond quickly. We were in Baghdad, engaging with Iraqis around the critical question of how technology can be used to help rebuild a society. As we moved around the city meeting with government ministers, military leaders, diplomats and Iraqi entrepreneurs, we encountered a nation whose prospects for recovery and future success appeared to hang by a thread. Eric's visit marked the first trip to Iraq by the CEO of a Fortune 500 technology company, so there were lots of questions about why Google was there. At the time, even we weren't entirely sure what Google might encounter or accomplish.

The answer became clear instantly. Everywhere we looked, we saw mobile devices. That surprised us. At the time, Iraq had been a war zone for more than six years, following the fall of Saddam Hussein, who, in his totalitarian paranoia, had banned the use of mobile phones. The war had decimated Iraq's physical infrastructure, and most people had unreliable access to food, water and electricity. Even basic commodities were prohibitively expensive. In some places, garbage hadn't been collected in *years*. And, critically, the security of the population was never guaranteed, either for high-level officials or for everyday shopkeepers. Mobile phones seemed like the last item that would appear on the country's dauntingly long to-do list. Yet as we came to learn, despite all of the pressing problems in their lives, Iraqis prioritized technology.

Not only did the Iraqis possess and value technology, they also saw its huge potential to improve their lives and the fate of their embattled country. The engineers and entrepreneurs we met expressed great frustration over their inability to help themselves. They already knew what they needed—reliable electricity, enough bandwidth for a fast connection, accessible digital tools and enough access to start-up capital to get their ideas off the ground.

It was Eric's first trip to a war zone, and Jared's umpteenth, yet we both came away with a sense that something profound was shifting in the world. If even war-weary Iraqis not only saw the possibilities of technology but knew what they wanted to do with it, how many other millions of people were out there with the drive and basic knowledge but not the access? For Jared, the trip confirmed to him that governments were dangerously behind the curve when it came to anticipating changes (fearful of them, too), and that they did not see the possibilities these new tools presented for tackling what challenges lay ahead. And Eric confirmed his feeling that the technology industry had many more problems to solve, and customers to serve, than anyone realized.

In the months following our trip, it became clear to us that there is a canyon dividing people who understand technology and people charged with addressing the world's toughest geopolitical issues, and no one has built a bridge. Yet the potential for collaboration between the tech industry, the public sector and civil society is enormous. As we thought about the spread of connectivity around the world, we found ourselves captivated by the questions generated by this divide: Who will be more powerful in the future, the citizen or the state? Will technology make terrorism easier or harder to carry out? What is the relationship between privacy and security, and how much will we have to give up to be part of the new digital age? How will war, diplomacy and revolution change when everyone is connected, and how can we tip the balance in a beneficial way? When broken societies are rebuilt, what will they be able to do with technology?

We collaborated first as writers of a memo to Secretary of State Hillary Clinton about lessons learned in Iraq, and thereafter as friends. We share a worldview about the potential of technology platforms, and their inherent power, and this informs all of the work we do, both within Google and outside it. We believe that modern technology platforms, such as Google, Facebook, Amazon and Apple, are even more powerful than most people realize, and our future world will be profoundly altered by their adoption and successfulness in societies everywhere. These platforms constitute a true paradigm shift, akin to the invention of television, and what gives them their power is their ability to grow—specifically, the speed at which they scale. Almost nothing short of a biological virus can spread as quickly, efficiently or aggressively as these technology platforms, and this makes the people who build, control and use them powerful too. Never before have so many people been connected through an instantly responsive network; the possibilities for collective action through communal online platforms (as consumers, creators, contributors, activists and in every other way) are truly game-changing. The scale effects that we're familiar with today, from a viral music video to an international e-commerce platform, merely hint at what is to come.

Because of digital platform-driven scale effects, things will happen much more quickly in the new digital age, with implications for every part of society, including politics, economics, the media, business and social norms. This acceleration to scale, when paired with the interconnectedness that Internet technology fosters, will usher in a new era of globalization—globalization of products *and* ideas. As members of the technology sector, it's our duty to fully and honestly explore the impact our industry's work has and will have on people's lives and on society, because, increasingly, governments will have to make rules synergistically with individuals and companies who are moving at an accelerated pace and pushing the boundaries sometimes faster than laws can keep up with. The digital platforms, networks and products they launch now have an outsized effect, on an international scale. So in order to understand the future of politics, business, diplomacy and other important sectors, one must understand how technology is driving major changes in those areas.

By coincidence, just as we began to share ideas about the future, a string of highly visible world events occurred that exemplified the very concepts and problems we were debating. The Chinese government launched sophisticated cyber attacks on Google and dozens of other American companies; WikiLeaks burst onto the scene, making hundreds of thousands of classified digital records universally accessible; major earthquakes in Haiti and Japan devastated cities but generated innovative tech-driven responses; and the revolutions of the Arab Spring shook the world with their speed, strength and contagious mobilization effects. Each turbulent development introduced new angles and possibilities about the future for us to consider.

We spent a great deal of time debating the meaning and consequences of events like these, predicting trends and theorizing possible tech-oriented solutions. This book is the product of those conversations.

In the forthcoming pages, we explore the future as we envision it, full of complex global issues involving citizenship, statecraft, privacy and war, among other issues, with both the challenges and the solutions driven by the rise of global connectivity. Where possible, we describe what can be done to help channel the influx of new technological tools in ways that inform, improve and enrich our world. Technology-driven change is inevitable, but at every stage, we can exert a measure of control over how it plays out. Some of the predictions you'll read in these pages will be things you've long suspected but couldn't admit—such as the logical conclusions of commercial drone warfare—while others will be wholly new. We hope that our predictions and recommendations will engage you and get you thinking.

This is not a book about gadgets, smart-phone apps or artificial intelligence, though each of these subjects will be discussed. This is a book about technology, but even more, it's a book about humans, and how humans interact with, implement, adapt to and exploit technologies in their environment, now and in the future, throughout the world. Most of all, this is a book about the importance of a guiding human hand in the new digital age. For all the possibilities that communication technologies represent, their use for good or ill depends solely on people. Forget all the talk about machines taking over. What happens in the future is up to us.

CHAPTER 1

Our Future Selves

Soon everyone on Earth will be connected. With five billion more people set to join the virtual world, the boom in digital connectivity will bring gains in productivity, health, education, quality of life and myriad other avenues in the physical world—and this will be true for everyone, from the most elite users to those at the base of the economic pyramid. But being "connected" will mean very different things to different people, largely because the problems they have to solve differ so dramatically. What might seem like a small jump forward for some—like a smart phone priced under $20—may be as profound for one group as commuting to work in a driverless car is for another. People will find that being connected virtually makes us feel more equal—with access to the same basic platforms, information and online resources—while significant differences persist in the physical world. Connectivity will not solve income inequality, though it will alleviate some of its more intractable causes, like lack of available education and economic opportunity. So we must recognize and celebrate innovation in its own context. Everyone will benefit from connectivity, but not equally, and how those differences manifest themselves in the daily lives of people is our focus here.

Increased Efficiency

Being able to do more in the virtual world will make the mechanics of our physical world more efficient. As digital connectivity reaches the far corners of the globe, new users will employ it to improve a wide range of inefficient markets, systems and behaviors, in both the most and least advanced societies. The resulting gains in efficiency and productivity will be profound, particularly in developing countries where technological isolation and bad policies have stymied growth and progress for years, and people will do more with less.

The accessibility of affordable smart devices, including phones and tablets, will be transformative in these countries. Consider the impact of basic mobile phones for a group of Congolese fisherwomen today. Whereas they used to bring their daily catch to the market and watch it slowly spoil as the day progressed, now they keep it on the line, in the river, and wait for calls from customers. Once an order is placed, a fish is brought out of the water and prepared for the buyer. There is no need for an expensive refrigerator, no need for someone to guard it at night, no danger of spoiled fish losing their value (or poisoning customers), and there is no unnecessary overfishing. The size of these women's market can even expand as other fishermen in surrounding areas coordinate with them over their own phones. As a substitute for a formal market economy (which would take years to develop), that's not a bad work-around for these women or the community at large.

Mobile phones are transforming how people in the developing world access and use information, and adoption rates are soaring. There are already more than 650 million mobile-phone users in Africa, and close to 3 billion across Asia. The majority of these people are using basic-feature phones—voice calls and text messages only—because the cost of data service in their countries is often prohibitively expensive, so that even those who can buy web-enabled phones or smart phones cannot use them affordably. This will change, and when it does, the smart-phone revolution will profoundly benefit these populations.

Hundreds of millions of people today are living the lives of their grandparents, in countries where life expectancy is less than sixty years, or even fifty in some places, and there is no guarantee that their political and macroeconomic circumstances will improve dramatically anytime soon. What is new in their lives and their futures is connectivity. Critically, they have the chance to bypass earlier technologies, like dial-up modems, and go directly to high-speed wireless connections, which means the transformations that connectivity brings will occur even more quickly than they did in the developed world. The introduction of mobile phones is far more transformative than most people in modern countries realize. As people come online, they will quite suddenly have access to almost all the world's information in one place in their own language. This will even be true for an illiterate Maasai cattle herder in the Serengeti, whose native tongue, Maa, is not written—he'll be able to verbally inquire about the day's market prices and crowd-source the whereabouts of any nearby predators, receiving a spoken answer from his device in reply. Mobile phones will allow formerly isolated people to connect with others very far away and very different from themselves. On the economic front, they'll find ways to use the new tools at their disposal to enlarge their businesses, make them more efficient and maximize their profits, as the fisherwomen did much more locally with their basic phones.

What connectivity also brings, beyond mobile phones, is the ability to collect and use data. Data itself is a tool, and in places where unreliable statistics about health, education, economics and the population's needs have stalled growth and development, the chance to gather data effectively is a game-changer. Everyone in society benefits from digital data, as governments can better measure the success of their programs, and media and other nongovernmental organizations can use data to support their work and check facts. For example, Amazon is able to take its data on merchants and, using algorithms, develop customized bank loans to offer them—in some cases when traditional banks have completely shut their doors. Larger markets and better metrics can help create healthier and more productive economies.

And the developing world will not be left out of the advances in gadgetry and other high-tech machinery. Even if the prices for sophisticated smart phones and robots to perform household tasks like vacuuming remain high, illicit markets like China's expansive *"shanzhai"* network for knock-off consumer electronics will produce and distribute imitations that bridge the gap. And technologies that emerged in first-world contexts will find renewed purpose in developing countries. In "additive manufacturing," or 3-D printing, machines can actually "print" physical objects by taking three-dimensional data about an object and tracing the contours of its shape, ultra-thin layer by ultra-thin layer, with liquid plastic or other material, until the whole object materializes. Such printers have produced a huge range of objects, including customized mobile phones, machine parts and a full-sized replica motorcycle. These machines will definitely have an impact on the developing world. Communal 3-D printers in poor countries would allow people to make whatever tool or item they require from open-source templates—digital information that is freely available in its edited source—rather than waiting on laborious or iffy delivery routes for higher-priced premade goods.

In wealthier countries 3-D printing will be the perfect partner for advanced manufacturing. New materials and products will all be built uniquely to a specification from the Internet and on demand by a machine run by a sophisticated, trained operator. This will not replace the acres of high-volume, lowest-cost manufacturing present in many industries, but it will bring an unprecedented variety to the products used in the developed world.

As for life's small daily tasks, information systems will streamline many of them for people living in those countries, such as integrated clothing machines (washing, drying, folding, pressing and sorting) that keep an inventory of clean clothes and algorithmically suggest outfits based on the user's daily schedule. Haircuts will finally be auto-mated and machine-precise. And cell phones, tablets and laptops will have wireless recharging capabilities, rendering the need to fiddle with charging cables an obsolete nuisance. Centralizing the many mov-ing parts of one's life into an easy-to-use, almost intuitive system of information management and decision making will give our interac-tions with technology an effortless feel. As long as safeguards are in place to protect privacy and prevent data loss, these systems will free us of many small burdens—including errands, to-do lists and assorted "monitoring" tasks—that today add stress and chip away at our men-tal focus throughout the day. Our own neurological limits, which lead us to forgetfulness and oversights, will be supplemented by informa-tion systems designed to support our needs. Two such examples are memory prosthetics—calendar reminders and to-do lists—and social prosthetics, which instantly connect you with your friend who has relevant expertise in whatever task you are facing.

By relying on these integrated systems, which will encompass both the professional and the personal sides of our lives, we'll be able to use our time more effectively each day—whether that means having the time to have a "deep think," spending more time preparing for an important presentation or guaranteeing that a parent can attend his or her child's soccer game without distraction. Suggestion engines that offer alternative terms to help a user find what she is looking for will be a particularly useful aid in efficiency by consistently stimulating our thinking processes, ultimately enhancing our creativity, not pre-empting it. Of course, the world will be filled with gadgets, holograms that allow a virtual version of you to be somewhere else, and endless amounts of content, so there will be plenty of ways to procrastinate, too—but the point is that when you choose to be productive, you can do so with greater capacity.

Other advances in the pipeline in areas like robotics, artificial intelligence and voice recognition will introduce efficiency into our lives by providing more seamless forms of engagement with the technology in our daily routines. Fully automated human-like robots with superb AI abilities will probably be out of most people's price range for some time, but the average American consumer will find it affordable to own a handful of different multipurpose robots fairly soon. The technology in iRobot's Roomba vacuum cleaner, the progenitor of this field of consumer "home" robots (first introduced in 2002), will only become more sophisticated and multipurpose in time. Future varieties of home robots should be able to handle other household duties, electrical work and even plumbing issues with relative ease.

We also can't discount the impact that superior voice-recognition software will have on our daily lives. Beyond searching for information online and issuing commands to your robots (both of which are possible today), better voice recognition will mean instant transcription of anything you produce: e-mails, notes, speeches, term papers. Most people speak much faster than they type, so this technology will surely save many of us time in our daily affairs—not to mention helping us avoid cases of carpal tunnel syndrome. A shift toward voice-initiated writing may well change our world of written material. Will we learn to speak in paragraphs, or will our writing begin to mirror speech patterns?

Everyday use of gesture-recognition technology is also closer than we think. Microsoft's Kinect, a hands-free sensor device for the

Xbox 360 video-game console that captures and integrates a player's motion, set a world record in 2011 as the fastest selling consumer-electronics device in history, with more than eight million devices sold in the first sixty days on the market. Gestural interfaces will soon move beyond gaming and entertainment into more functional areas; the futuristic information screens displayed so prominently in the film *Minority Report*—in which Tom Cruise used gesture technology and holographic images to solve crimes on a computer—are just the beginning. In fact, we've already moved beyond that—the really interesting work today is building "social robots" that can recognize human gestures and respond to them in kind, such as a toy dog that sits when a child makes a command gesture.

And, looking further down the line, we might not need to move physically to manipulate those robots. There have been a series of exciting breakthroughs in thought-controlled motion technology—directing motion by thinking alone—in the past few years. In 2012, a team at a robotics laboratory in Japan demonstrated successfully that a person lying in an fMRI machine (which takes continuous scans of the brain to measure changes in blood flow) could control a robot hundreds of miles away just by imagining moving different parts of his body. The subject could see from the robot's perspective, thanks to a camera on its head, and when he thought about moving his arm or his legs, the robot would move correspondingly almost instantaneously. The possibilities of thought-controlled motion, not only for "surrogates" like separate robots but also for prosthetic limbs, are particularly exciting in what they portend for mobility-challenged or "locked in" individuals—spinal-cord-injury patients, amputees and others who cannot communicate or move in their current physical state.

More Innovation, More Opportunity

That the steady march of globalization will continue apace, even accelerate, as connectivity spreads will come as no surprise. But what might surprise you is how small some of the advances in technology, when paired with increased connection and interdependence across countries, will make your world feel. Instant language translation, virtual-reality interactions and real-time collective editing—most easily understood today as wikis—will reshape how firms and organizations interact with partners, clients and employees in other places. While certain differences will perhaps never be fully overcome—like cultural nuance and time zones—the ability to engage with people in disparate locations, with near-total comprehension and on shared platforms, will make such interactions feel incredibly familiar.

Supply chains for corporations and other organizations will become increasingly disaggregated, not just on the production side but also with respect to people. More effective communication across borders and languages will build trust and create opportunities for hardworking and talented individuals around the world. It will not be unusual for a French technology company to operate its sales team from Southeast Asia, while locating its human-resources people in Canada and its engineers in Israel. Bureaucratic obstacles that prevent this level of decentralized operation today, like visa restrictions and regulations around money transfers, will become either irrelevant or be circumvented as digital solutions are discovered. Perhaps a human-rights organization with staff living in a country under heavy diplomatic sanctions will pay its employees in mobile money credits, or in an entirely digital currency.

As fewer jobs require a physical presence, talented individuals will have more options available to them. Skilled young adults in Uruguay will find themselves competing for certain types of jobs against their counterparts in Orange County. Of course, just as not all jobs can or will be automated in the future, not every job can be conducted from a distance—but more can than you might think. And for those living on a few dollars per day, there will be endless opportunities to increase their earnings. In fact, Amazon Mechanical Turk, which is a digital task-distribution platform, offers a present-day example of a company outsourcing small tasks that can be performed for a few cents by anyone with an Internet connection. As the quality of virtual interactions continues to improve, a range of vocations can expand the platform's client base; you might retain a lawyer from one continent and use a Realtor from another. Globalization's critics will decry this erosion of local monopolies, but it should be embraced, because this is how our societies will move forward and continue to innovate. Indeed, rising connectivity should *help* countries discover their competitive advantage—it could be that the world's best graphic designers come from Botswana, and the world just doesn't know it yet.

This leveling of the playing field for talent extends to the world of ideas, and innovation will increasingly come from the margins, outside traditional bastions of growth, as people begin to make new connections and apply unique perspectives to difficult problems, driving change. New levels of collaboration and cross-pollination across different sectors internationally will ensure that many of the best ideas and solutions will have a chance to rise to the top and be seen, considered, explored, funded, adopted and celebrated. Perhaps an aspiring Russian programmer currently working as a teacher in Novosibirsk will discover a new application of the technology behind the popular mobile game Angry Birds, realizing how its game framework could be used to improve the educational tools he is building to teach physics to local students. He finds similar gaming software that is open source and then he builds on it. As the open-source movement around the world continues to gain speed (for governments and companies it is low cost, and for contributors the benefits are in recognition and economic opportunities to improve and enlarge the support ecosystems), the Russian teacher-programmer will have an enormous cache of technical plans to learn from and use in his own work. In a fully connected world, he is increasingly likely to catch the eyes of the right people, to be offered jobs or fellowships, or to sell his creation to a major multinational company. At a minimum, he can get his foot in the door.

Innovation can come from the ground up, but not all local innovation will work on a larger scale, because some entrepreneurs and inventors will be building for different audiences, solving very specific problems. This is true today as well. Consider the twenty-four-year-old Kenyan inventor Anthony Mutua, who unveiled at a 2012 Nairobi science fair an ultrathin crystal chip he developed that can generate electricity when put under pressure. He placed the chip in the sole of a tennis shoe and demonstrated how, just by walking, a person can charge his mobile phone. (It's a reminder of how bad the problems of reliable and affordable electricity, and to a lesser extent short battery life, are for many people—and how some governments are not rushing to fix the electricity grids—that innovators like Mutua are designing microchips that turn people into portable charging stations.) Mutua's chip is now set to go into mass production, and if that successfully brings down the cost, he will have invented one of the cleverest designs that no one outside the developing world will ever use, simply because they'll never need to. Unfortunately, the level of a population's access to technology is often determined by external factors, and even if power and electricity problems are eventually solved (by the government or by citizens), there is no telling what new roadblocks will prevent certain groups from reaching the same level of connectivity and opportunity as others.

The most important pillar behind innovation and opportunity—education—will see tremendous positive change in the coming decades as rising connectivity reshapes traditional routines and offers new paths for learning. Most students will be highly technologically literate, as schools continue to integrate technology into lesson plans and, in some cases, replace traditional lessons with more interactive workshops. Education will be a more flexible experience, adapting itself to children's learning styles and pace instead of the other way around. Kids will still go to physical schools, to socialize and be guided by teachers, but as much, if not more, learning will take place employing carefully designed educational tools in the spirit of today's Khan Academy, a nonprofit organization that produces thousands of short videos (the majority in science and math) and shares them online for free. With hundreds of millions of views on the Khan Academy's YouTube channel already, educators in the United States are increasingly adopting its materials and integrating the approach of its founder, Salman Khan—modular learning tailored to a student's needs. Some are even "flipping" their classrooms, replacing lectures with videos watched at home (as homework) and using school time for traditional homework, such as filling out a problem set for math class. Critical thinking and problem-solving skills will become the focus in many school systems as ubiquitous digital-knowledge tools, like the more accurate sections of Wikipedia, reduce the importance of rote memorization.

For children in poor countries, future connectivity promises new access to educational tools, though clearly not at the level described above. Physical classrooms will remain dilapidated; teachers will continue to take paychecks and not show up for class; and books and supplies will still be scarce. But what's new in this equation—connectivity—promises that kids with access to mobile devices and the Internet will be able to experience school physically *and* virtually, even if the latter is informal and on their own time.

In places where basic needs are poorly met by the government, or in insecure areas, basic digital technologies like mobile phones will offer safe and inexpensive options for families looking to educate their children. A child who cannot attend school due to distance, lack of security or school fees will have a lifeline to the world of learning if she has access to a mobile phone. Even for those children without access to data plans or the mobile web, basic mobile services, like text messages and IVR (interactive voice response, a form of voice-recognition technology), can provide educational outlets. Loading tablets and mobile phones with high-quality education applications and entertainment content before they are sold will ensure that the "bandwidth poor," who lack reliable connectivity, will still benefit from access to these devices. And for children whose classrooms are overcrowded or understaffed, or whose national curriculum is dubiously narrow, connectivity through mobile devices will supplement their education and help them reach their full potential, regardless of their origins. Today numerous pilot projects exist in developing countries that leverage mobile technology to teach a wide range of topics and skills, including basic literacy for children and adults, second languages and advanced courses from universities. In 2012, the MIT Media Lab tested this approach in Ethiopia by distributing preloaded tablets to primary-age kids without instructions or accompanying teachers. The results were extraordinary: within months the kids were reciting the entire alphabet and writing complete sentences in English. Without the connectivity that will be ubiquitous in the future, there are limits to what any of these efforts can accomplish today.

Just imagine the implications of these burgeoning mobile or tablet-based learning platforms for a country like Afghanistan, which has one of the lowest rates of literacy in the world. Digital platforms, whether presented in simple mobile form or in more sophisticated ways online, will eventually be able to withstand any environmental turbulence (political instability, economic collapse, perhaps even bad weather) and continue to serve the needs of users. So while the educational experience in the physical world will remain volatile for many, the virtual experience will increasingly become the more important and predictable option. And students stuck in school systems that teach narrow curriculums or only rote memorization will have access to a virtual world that encourages independent exploration and critical thinking.

A Better Quality of Life

In tandem with the wide variety of functional improvements in your daily life, future connectivity promises a dazzling array of "quality of life" improvements: things that make you healthier, safer and more engaged. As with other gains, there remains a sliding scale of access here, but that doesn't make them any less meaningful.

The devices, screens and various machines in your future apartment will serve a purpose beyond utility—they will offer entertainment, wanted distraction, intellectual and cultural enrichment, relaxation and opportunities to share things with others. The key advance ahead is personalization. You'll be able to customize your devices—indeed, much of the technology around you—to fit your needs, so that your environment reflects your preferences. People will have a better way to curate their life stories and will no longer have to rely on physical or online photo albums, although both will still exist. Future videography and photography will allow you to project any still or moving image you've captured as a three-dimensional holograph. Even more remarkable, you will be able to integrate any photos, videos and geographic settings that you choose to save into a single holographic device that you will place on the floor of your living room, instantaneously transforming the space into a memory room. A couple will be able to re-create their wedding ceremony for grandparents who were too ill to attend.

What you can watch on your various displays (high-quality LCD—liquid crystal display—screens, holographic projections or a handheld mobile device) will be determined by you, not by network-television schedules. At your fingertips will be an entire world's worth of digital content, constantly updated, ranked and categorized to help you find the music, movies, shows, books, magazines, blogs and art you like. Individual agency over entertainment and information channels will be greater than ever, as content producers shift from balkanized protectiveness to more unified and open models, since a different business model will be necessary in order to keep the audience. Contemporary services like Spotify, which offers a large catalog of live-streaming music for free, give us a sense of what the future will look like: an endless amount of content, available anytime, on almost any device, and at little or no cost to users, with copyrights and revenue streams preserved. Long-standing barriers to entry for content creators are being flattened as well; just as YouTube can be said to launch careers today[*] (or at least offer fleeting fame), in the future, even more platforms will offer artists, writers, directors, musicians and others in every country the chance to reach a wider audience. It will still require skill to create quality content, but it will also be easier to assemble a team with the requisite skills to do this—say, an animator from South Korea, a voice actor from the Philippines, a storyboarder from Mexico and a musician from Kenya—and the finished product may have the potential to reach as wide an audience as any Hollywood blockbuster.

Entertainment will become a more immersive and personalized experience in the future. Integrated tie-ins will make today's product placements seem passive and even clumsy. If while watching a television show you spot a sweater you want or a dish you think you'd like to cook, information including recipes or purchasing details will be readily available, as will every other fact about the show, its story lines, actors and locations. If you're feeling bored and want to take an hour-long holiday, why not turn on your holograph box and visit Carnival in Rio? Stressed? Go spend some time on a beach in the Maldives. Worried your kids are becoming spoiled? Have them spend some time wandering around the Dharavi slum in Mumbai. Frustrated by the media's coverage of the Olympics in a different time zone? Purchase a holographic pass for a reasonable price and watch the women's gymnastics team compete right in front of you, live. Through virtual-reality interfaces and holographic-projection capabilities, you'll be able to "join" these activities as they happen and experience them as if you were truly there. Nothing beats the real thing, but this will be a very

---

[*] The Korean K-pop star Psy's fame reached global proportions almost overnight as the video he created for his song "Gangnam Style" became the most-watched YouTube video ever within a span of three months.

close second. And if nothing else, it will certainly be more affordable. Thanks to these new technologies, you can be more stimulated, or more relaxed, than ever before.

You'll be safer, too, at least on the road. While some of the very exciting new possibilities in transportation, like supersonic tube commutes and suborbital space travel, are still far in the distance, ubiquitous self-driving cars are imminent. Google's fleet of driverless cars, built by a team of Google and Stanford University engineers, has logged hundreds of thousands of miles without incident, and other models will soon join it on the road. Rather than replacing drivers altogether, the liminal step will be a "driver-assist" approach, where the self-driving option can be turned on, just as an airline captain turns on the autopilot. Government authorities are already well versed on self-driving cars and their potential—in 2012, Nevada became the first state to issue licenses to driverless cars, and later that same year California also affirmed their legality. Imagine the possibilities for long-haul truck-driving. Rather than testing the biological limits of human drivers with thirty-hour trips, the computer can take over primary responsibility and drive the truck for stretches as the driver rests.

The advances in health and medicine in our near future will be among the most significant of all the new game-changing developments. And thanks to rising connectivity, an even wider range of people will benefit than at any other time in history. Improvements in disease detection and treatment, the management of medical records and personal-health monitoring promise more equitable access to health care and health information for potentially billions more people when we factor in the spread of digital technology.

The diagnostic capability of your mobile phone will be old news. (*Of course* you will be able to scan body parts the way you do bar codes.) But soon you will be benefiting from a slew of physical augmentations designed to monitor your well-being, such as microscopic robots in your circulatory system that keep track of your blood pressure, detect nascent heart disease and identify early-stage cancer. Inside your grandfather's new titanium hip there will be a chip that can act as a pedometer, monitor his insulin levels to check for the early stages of diabetes, and even trigger an automated phone call to an emergency contact if he takes a particularly hard fall and might need assistance. A tiny nasal implant will be available to you that will alert you to airborne toxins and early signs of a cold.

Eventually these accoutrements will be as uncontroversial as artificial pacemakers (the first of which was implanted in the 1950s). They are the logical extensions of today's personal-health-tracking applications, which allow people to use their smart phones to log their exercise, track their metabolic rates and chart their cholesterol levels. Indeed, ingestible health technology already exists—the Food and Drug Administration (FDA) approved the first electronic pill in 2012. Made by a California-based biomedical firm called Proteus Digital Health, the pill carries a tiny sensor one square millimeter in size, and once the pill is swallowed, stomach acid activates the circuit and sends a signal to a small patch worn outside the body (which then sends its data to a mobile phone). The patch can collect information about a patient's response to a drug (monitoring body temperature, heart rate and other indicators), relay data about regular usage to doctors and even track what a person eats. For sufferers of chronic illnesses and the elderly particularly, this technology will allow for significant improvements: automatic reminders to take various medications, the ability to measure directly how drugs are reacting in a person's body and the creation of an instant digital feedback loop with doctors that is personalized and data-driven. Not everyone will want to actively oversee their health to this degree, let alone the even more detailed version of the future, but they probably will want their doctor to have access to such data. "Intelligent pills" and nasal implants will be sufficiently affordable so as to be as accessible as vitamins and supplements. In short order, we will have access to personal health-care systems run off of our mobile devices that will automatically detect if something is wrong with us based on data collected from some of the above-mentioned augmentations, prompt us with appointment options for a nearby doctor and subsequently (with consent) send all of the relevant data about our symptoms and health indicators to the doctor being consulted.

Tissue engineers will be able to grow new organs to replace patients' old or diseased ones, using either synthetic materials or a person's own cells. At the outset, affordability will limit the use. Synthetic skin grafts, which exist today, will give way to grafts made from burn victims' own cells. Inside hospitals, robots will take on more responsibilities, as surgeons increasingly let sophisticated machines handle difficult parts of certain procedures, where delicate or tedious work is involved or a wider range of motion is required.*

Advances in genetic testing will usher in the era of personalized medicine. Through targeted tests and genome sequencing (decoding a person's full DNA), doctors and disease specialists will have more information about patients, and what might help them, than ever before. Despite steady scientific progress, severe negative reactions to prescribed drugs remain a leading cause of hospitalization and death. Pharmaceutical companies traditionally pursue a "one-size-fits-all" approach to drug development, but this is due to change as the burgeoning field of pharmacogenetics continues to develop. Better genetic testing will reduce the likelihood of negative reactions, improve patients' chances and provide doctors and medical researchers with more data to analyze and use. Eventually, and initially only for the wealthy, it will be possible to design pharmaceutical drugs tailored to an individual's genetic structure. But this too will change as the cost of DNA sequencing drops below $100 and almost everything biological is sequenced, making it possible for a much broader segment of the world's population to benefit from highly specific, personalized diagnoses.

For those living in developing countries, basic connectivity and access to the virtual world will offer a resource they can leverage to improve their own quality of life, and nowhere more so than in the area of health. Even though their environment in the physical world is colored by inadequate care, lack of available vaccines and medicines, broken health systems and other exogenous factors that create health crises (like conflict-related internal migration), many important gains in health care will be driven by innovative uses of mobile phones, largely by individuals and other nongovernmental actors who seize the opportunity to drive change in an otherwise stagnant system. We already see this happening. Across the developing world today, the "mobile health" revolution—mobile phones used as tools to connect patients to doctors, to monitor drug distribution and to increase

---

* Robotic surgical suites are already in operation in hospitals in the United States and Europe.

the reach of health clinics—is responsible for a number of improvements as a range of technology start-ups, nonprofits and entrepreneurs tackle difficult problems with technology-first solutions. Mobile phones are now used to track drug shipments and verify their authenticity, to share basic health information that isn't available locally, to send reminders about medication and appointments to patients, and to gather data about health indicators that government officials, NGOs and other actors can use to design their programs. The central problems in health sectors in poor places, like understaffed clinics, underserved patients in remote places, too few medications or inefficient distribution of them, and misinformation about vaccines and disease prevention, will all find at least partial solutions through connectivity.

At the very least, the adoption of mobile phones gives people a new level of agency over their personal health, even though the devices themselves, of course, can't cure illness. People can use their phones to access information about preventative health care or recovery. They can use basic diagnostic tools embedded in their phones—maybe not X-rays, but cameras and audio recordings. A woman can take a picture of a lesion, or a recording of a cough, and send that information to a doctor or health professional, whom she can then interact with remotely, efficiently, affordably and privately. Digital solutions like these are not a perfect substitute for a properly functioning health sector, but in the meantime, they can offer new information and interactions that at a minimum will chip away at a larger and more entrenched multigenerational problem.

The Upper Band

Connectivity benefits everyone. Those who have none will have some, and those who have a lot will have even more. To demonstrate that, imagine you are a young urban professional living in an American city a few decades from now. An average morning might look something like this:

There will be no alarm clock in your wake-up routine—at least, not in the traditional sense. Instead, you'll be roused by the aroma of freshly brewed coffee, by light entering your room as curtains open automatically, and by a gentle back massage administered by your high-tech bed. You're more likely to awake refreshed, because inside your mattress there's a special sensor that monitors your sleeping rhythms, determining precisely when to wake you so as not to interrupt a REM cycle.

Your apartment is an electronic orchestra, and you are the conductor. With simple flicks of the wrist and spoken instructions, you can control temperature, humidity, ambient music and lighting. You are able to skim through the day's news on translucent screens while a freshly cleaned suit is retrieved from your automated closet because your calendar indicates an important meeting today. You head to the kitchen for breakfast and the translucent news display follows, as a projected hologram hovering just in front of you, using motion detection, as you walk down the hallway. You grab a mug of coffee and a fresh pastry, cooked to perfection in your humidity-controlled oven—and skim new e-mails on a holographic "tablet" projected in front of you. Your central computer system suggests a list of chores your housekeeping robots should tackle today, all of which you approve. It further suggests that, since your coffee supply is projected to run out next Wednesday, you consider purchasing a certain larger-size container that it noticed currently on sale online. Alternatively, it offers a few recent reviews of other coffee blends your friends enjoy.

As you mull this over, you pull up your notes for a presentation you'll give later that day to important new clients abroad. All of your data—from your personal and professional life—is accessible through all of your various devices, as it's stored in the cloud, a remote digital-storage system with near limitless capacity. You own a few different and interchangeable digital devices; one is the size of a tablet, another the size of a pocket watch, while others might be flexible or wearable. All will be lightweight, incredibly fast and will use more powerful processors than anything available today.

You take another sip of coffee, feeling confident that you'll impress your clients. You already feel as if you know them, though you've never met in person, since your meetings have been conducted in a virtual-reality interface. You interact with holographic "avatars" that exactly capture your clients' movements and speech. You understand them and their needs well, not least because autonomous language-translation software reproduces the speech of both parties in perfect translations almost instantly. Real-time virtual interactions like these, as well as the ability to edit and collaborate on documents and other projects, makes the actual distance between you seem negligible.

As you move about your kitchen, you stub your toe, hard, on the edge of a cabinet—ouch! You grab your mobile device and open the diagnostics app. Inside your device there is a tiny microchip that uses low-radiation submillimeter waves to scan your body, like an X-ray. A quick scan reveals that your toe is just bruised, not broken. You decline the invitation your device suggests to get a second opinion at a nearby doctor's office.

There's a bit of time left before you need to leave for work—which you'll get to by driverless car, of course. Your car knows what time you need to be in the office each morning based on your calendar and, after factoring in traffic data, it communicates with your wristwatch to give you a sixty-minute countdown to when you need to leave the house. Your commute will be as productive or relaxing as you desire.

Before you head out, your device reminds you to buy a gift for your nephew's upcoming birthday. You scan the system's proposed gift ideas, derived from anonymous, aggregated data on other nine-year-old boys with his profile and interests, but none of the suggestions inspire you. Then you remember a story his parents told you that had everyone forty and older laughing: Your nephew hadn't understood a reference to the old excuse "A dog ate my homework"; how could a dog eat his cloud storage drive? He had never gone to school before digital textbooks and online lesson plans, and he had used paper to do his homework so rarely—and used cloud storage so routinely—that the notion that he would somehow "forget" his homework *and* come up with an excuse like that struck him as absurd. You do a quick search for a robotic dog and buy one with a single click, after adding a few special touches he might like, such as a reinforced titanium skeleton so that he can ride on it. In the card input, you type: "Just in case." It will arrive at his house within a five-minute window of your selected delivery time.

You think about having another cup of coffee, but then a haptic device ("haptic" refers to technology that involves touch and feeling) that is embedded in the heel of your shoe gives you a gentle pinch—a signal that you'll be late for your morning meeting if you linger any longer. Perhaps you grab an apple on the way out, to eat in the backseat of your car as it chauffeurs you to your office.

If you are a part of the world's upper band of income earners (as most residents of wealthy Western countries are), you will have access to many of these new technologies directly, as owners or as friends of those who own them. You probably recognize from this morning routine a few things you have already imagined or experienced. Of course, there will always be the super-wealthy people whose access to technology will be even greater—they'll probably eschew cars altogether and travel to work in motion-stabilized automated helicopters, for example.

We will continue to encounter challenges in the physical world, but the expansion of the virtual world and what is possible online—as well as the inclusion of five billion more minds—means we will have new ways of getting information and moving resources to solve those problems, even if the solutions are imperfect. While there will remain significant differences between us, more opportunities to interact and better policy can help blur the edges.

The advance of connectivity will have an impact far beyond the personal level; the ways that the physical and virtual worlds coexist, collide and complement each other will greatly affect how citizens and states behave in the coming decades. And not all the news is good. The coming chapters delve into how everyone—individuals, companies, nongovernmental organizations (NGOs), governments and others—will handle this new reality of existing in both worlds, and how they will leverage the best and worst of what each world has to offer in the new digital age. Each individual, state and organization will have to discover its own formula, and those that can best navigate this multi-dimensional world will find themselves ahead in the future.

CHAPTER 2

The Future of Identity, Citizenship and Reporting

In the next decade, the world's virtual population will outnumber the population of Earth. Practically every person will be represented in multiple ways online, creating vibrant and active communities of interlocking interests that reflect and enrich our world. All of those connections will create massive amounts of data—a data revolution, some call it—and empower citizens in ways never before imagined. Yet despite these advancements, a central and singular caveat exists: The impact of this data revolution will be to strip citizens of much of their control over their personal information in virtual space, and that will have significant consequences in the physical world. This may not be true in every instance or for every user, but on a macro level it will deeply affect and shape our world. The challenge we face as individuals is determining what steps we are willing to take to regain control over our privacy and security.

Today, our online identities affect but rarely overshadow our physical selves. What people do and say on their social-networking profiles can draw praise or scrutiny, but for the most part truly sensitive or personal information stays hidden from public view. Smear campaigns and online feuds typically involve public figures, not ordinary citizens. In the future, our identities in everyday life will come to be defined more and more by our virtual activities and associations. Our highly documented pasts will have an impact on our prospects, and our ability to influence and control how we are perceived by others will decrease dramatically. The potential for someone else to access, share or manipulate parts of our online identities will increase, particularly due to our reliance on cloud-based data storage. (In nontechnical language, cloud computing refers to software hosted on the Internet that the user does not need to closely manage. Storing documents or content "in the cloud" means that data is stored on remote servers rather than on local ones or on a person's own computer, and it can be accessed by multiple networks and users. With cloud computing, online activities are faster, quicker to spread and better equipped to handle traffic loads.) This vulnerability—both perceived and real—will mandate that technology companies work even harder to earn the trust of their users. If they do not exceed expectations in terms of both privacy and security, the result will be either a backlash or abandonment of their product. The technology industry is already hard at work to find creative ways to mitigate risks, such as through two-factor authentication, which requires you to provide two of the following to access your personal data: something you know (e.g., password), have (e.g., mobile device) and are (e.g., thumbprint). We are also encouraged knowing that many of the world's best engineers are hard at work on the next set of solutions. And at a minimum, strong encryption will be nearly universally adopted as a better but not perfect solution. ("Encryption" refers to the scrambling of information so that it can be decoded and used only by someone with the right verification requirements.)

The basics of online identity could also change. Some governments will consider it too risky to have thousands of anonymous, untraceable and unverified citizens—"hidden people"; they'll want to know who is associated with each online account, and will require verification, at a state level, in order to exert control over the virtual world. Your online identity in the future is unlikely to be a simple Facebook page; instead it will be a constellation of profiles, from every online activity, that will be verified and perhaps even regulated by the government. Imagine all of your accounts—Facebook, Twitter, Skype, Google+, Netflix, *New York Times* subscription—linked to an "official profile." Within search results, information tied to verified online profiles will be ranked higher than content without such verification, which will result in most users naturally clicking on the top (verified) results. The true cost of remaining anonymous, then, might be irrelevance; even the most fascinating content, if tied to an anonymous profile, simply won't be seen because of its excessively low ranking.

The shift from having one's identity shaped off-line and projected online to an identity that is fashioned online and experienced off-line will have implications for citizens, states and companies as they navigate the new digital world. And how people and institutions handle privacy and security concerns in this formative period will determine the new boundaries for citizens everywhere. We want to explore here what full connectivity will mean for citizens in the future, how they will react to it and what consequences it will have for dictators and democrats alike.

The Data Revolution

The data revolution will bring untold benefits to the citizens of the future. They will have unprecedented insight into how other people think, behave and adhere to norms or deviate from them, both at home and in every society in the world. The newfound ability to obtain accurate and verified information online, easily, in native languages and in endless quantity, will usher in an era of critical thinking in societies around the world that before had been culturally isolated. In societies where the physical infrastructure is weak, connectivity will enable people to build businesses, engage in online commerce and interact with their government at an entirely new level.

The future will usher in an unprecedented era of choices and options. While some citizens will attempt to manage their identity by engaging in the minimum amount of virtual participation, others will find the opportunities to participate worth the risk of the exposure they incur. Citizen participation will reach an all-time high as anyone with a mobile handset and access to the Internet will be able to play a part in promoting accountability and transparency. A shopkeeper in Addis Ababa and a precocious teenager in San Salvador will be able to disseminate information about bribes and corruption, report election irregularities and generally hold their governments to account. Video cameras installed in police cars will help keep the police honest, if the camera phones carried by citizens don't already. In fact, technology will empower people to police the police in a plethora of creative ways never before possible, including through real-time monitoring systems allowing citizens to publicly rate every police officer in their hometown. Commerce, education, health care and the justice system will all become more efficient, transparent and inclusive as major institutions opt in to the digital age.

People who try to perpetuate myths about religion, culture, ethnicity or anything else will struggle to keep their narratives afloat amid a sea of newly informed listeners. With more data, everyone gains a better frame of reference. A Malawian witch doctor might find his community suddenly hostile if enough people find and believe information online that contradicts his authority. Young people in Yemen might confront their tribal elders over the traditional practice of child brides if they determine that the broad consensus of online voices is against it, and thus it reflects poorly upon them personally. Or followers of an Indian holy man might find a way to cross-reference his credentials on the Internet, abandoning him if it is revealed that he misled them. While many worry about the phenomenon of confirmation bias (when consciously or otherwise, people pay attention to sources of information that reinforce their existing worldview) as online sources of information proliferate, a recent Ohio State University study suggests that this effect is weaker than perceived, at least in the American political landscape. In fact, confirmation bias is as much about our responses to information passively received as it is about our tendency to proactively select information sources. So as millions of people come online we have reason to be optimistic about the social changes ahead.

Governments, too, will find it more difficult to maneuver as their citizens become more connected. Destroying documents, kidnapping, demolishing monuments—restrictive and repressive actions like these will lose much of their functional and symbolic power in the new digital age. Those documents would be recoverable, having been stored in the cloud, and the pressure that an active and globalized Internet community can produce when rallied against injustice will make governments think twice before snatching anyone or detaining him indefinitely. A Taliban-like government would still be able to destroy monuments like the Bamiyan Buddhas, but in the future those monuments will have been scanned with sophisticated technology that preserves every nook and cranny in virtual memory, allowing them to be rebuilt later by men or 3-D printers, or even projected as a hologram. Perhaps the UNESCO World Heritage Centre will add these practices to its restoration efforts. The structure of Syria's oldest synagogue, for example, currently in a museum in Damascus, could be projected as a hologram or reconstructed using 3-D printing at its original site in Dura-Europos. What's true now in most developed countries—the presence of an active civil society keen to fact-check and investigate its government—will be true almost everywhere, aided significantly by the prevalence of cheap and powerful handsets. And on a more basic level, citizens anywhere will be able to compare themselves and their way of life with the rest of the world. Practices widely considered barbaric or backward will seem even more so when seen in that context.

Identity will be the most valuable commodity for citizens in the future, and it will exist primarily online. Online experience will start with birth, or even earlier. Periods of people's lives will be frozen in time, and easily surfaced for all to see. In response, companies will have to create new tools for control of information, such as lists that would enable people to manage who sees their data. The communication technologies we use today are invasive by design, collecting our photos, comments and friends into giant databases that are searchable and, in the absence of outside regulation, fair game for employers, university admissions personnel and town gossips. We are what we tweet.

Ideally, all people would have the self-awareness to closely manage their online identities and the virtual lives they lead, monitoring and shaping them from an early age so as not to limit their opportunities in life. Of course, this is impossible. For children and adolescents, the incentives to share will always outweigh the vague, distant risks of self-exposure, even with salient examples of the consequences in public view. By the time a man is in his forties, he will have accumulated and stored a comprehensive online narrative, all facts and fictions, every misstep and every triumph, spanning every phase of his life. Even the rumors will live forever.

In deeply conservative societies where social shame is weighed heavily, we could see a kind of "virtual honor killing"—dedicated efforts to ruin a person's online identity either preemptively (by exposing perceived misdeeds or planting false information) or reactively (by linking his or her online identity to content detailing a crime, real or imagined). Ruined online reputations might not lead to physical violence by the perpetrator, but a young woman facing such accusations could find herself branded with a digital scarlet letter that, thanks to the unfortunate but hard-to-prevent reality of data permanence, she'd never be able to escape. And that public shame could lead one of her family members to kill her.

And what about the role of parents? Being a parent is hard enough, as anyone who has kids knows. While the online world has made it even tougher, it is not a hopeless endeavor. Parents will have the same responsibilities in the future, but they will need to be even more involved if they are going to make sure their children do not make mistakes online that could hurt their physical future. As children live significantly faster lives online than their physical maturity allows, most parents will realize that the most valuable way to help their child is to have the privacy-and-security talk even before the sex talk. The old-fashioned tactic of parents talking to their children will retain enormous value.

School systems will also adapt to play an important role. Parent-teacher associations will advocate for privacy and security classes to be taught alongside sex-education classes in their children's schools. Such classes will teach students to optimize their privacy-and-security settings and train them to become well versed in the dos and don'ts of the virtual world. And teachers will frighten them with real-life stories of what happens if they don't take control of their privacy and security at an early age.

Certainly some parents will try to game the system as well with more algorithmic solutions that may or may not have an effect. The process of naming a child offers one such example. As the functional value of online identity increases, parental supervision will play a critical role in the early stages of life, beginning with a child's name. Steven D. Levitt and Stephen J. Dubner, the authors of the popular economics book *Freakonomics,* famously dissected how ethnically popular names (specifically, names common in African-American communities) can be an indicator of children's chances for success in life. Looking ahead, parents will also consider how online search rankings will affect their child's future. The truly strategic will go beyond reserving social-networking profiles and buying domain names (e.g., www .JohnDavidSmith.com), and instead select names that affect how easy or hard it will be to find their children online. Some parents will deliberately choose unique names or unusually spelled traditional names so that their children have an edge in search results, making them easy to locate and promotable online without much direct competition. Others will go the opposite route, choosing basic and popular names that allow their children to live in an online world with some degree of shelter from Internet indexes—just one more "Jane Jones" among thousands of similar entries.

We'll also see a proliferation of businesses that cater to privacy and reputation concerns. This industry exists already, with companies like Reputation.com using a range of proactive and reactive tactics to remove or dilute unwanted content from the Internet.* During the 2008 economic crash, it was reported that several Wall Street bankers hired online reputation companies to minimize their appearance online, paying up to $10,000 per month for the service. In the future, this industry will diversify as the demand explodes, with identity managers becoming as common as stockbrokers and financial planners. Active management of one's online presence—say, by receiving quarterly reports from your identity manager tracking the changing shape of your online identity—will become the new normal for the prominent and those who aspire to be prominent.

A new realm of insurance will emerge, too. Companies will offer to insure your online identity against theft and hacking, fraudulent accusations, misuse or appropriation. For example, parents may take out an insurance policy against reputational damage caused by what their children do online. Perhaps a teacher will take out an insurance policy that covers her against a student hacking into her Facebook account and changing details of her online profile to embarrass or defame her. We have identity-theft protection companies today; in the future, insurance companies will offer customers protection against very spe-

---

* Most of these techniques fall under the umbrella of search-engine optimization (SEO) processes. To influence the ranking algorithm of search engines, the most common method is to seed positive content around the target (e.g., a person's name), encourage links to it and frequently update it, so that the search-engine spiders are likely to identify the material as popular and new, which pushes down the older, less relevant content. Using prominent keywords and adding back-links (incoming links to a website) to popular sites can also influence the ranking. This is all legal and generally considered fair. There is an underside to SEO, however—"black-hat SEO"—where efforts to manipulate rankings include less legal or fair practices like sabotaging other content (by linking it to red-flag sites like child pornography), adding hidden text or cloaking (tricking the spiders so that they see one version of the site while the end user sees another).

cific misuses. Any number of people could be attracted to such an insurance policy, from the genuinely in need to the generally paranoid.

Online identity will become such a powerful currency that we will even see the rise of a new black market where people can buy real or invented identities. Citizens and criminals alike will be attracted to such a network, since the false identity that could provide cover for a known drug smuggler could also shelter a political dissident. The identity will be manufactured or stolen, and it will come complete with backdated entries and IP (Internet protocol) activity logs, false friends and sales purchases, and other means of making it appear convincing. If a Mexican whistle-blower's family needed to flee the violence of Ciudad Juárez and feared cartel retribution, a set of fake online identities would certainly help cover their tracks and provide them with a clean slate.

Naturally, this kind of escape route is a high-risk endeavor in the digital age: Embarking on a new life would require total disconnection from previous ties, because even the smallest gesture (like a search query for a relative) could give away a person's position. Furthermore, anyone assuming a false identity would need to avoid all places with facial-recognition technology lest a scan of his or her face flag an earlier profile. And there would be no dark alleyways in this illicit market, either: All identities could be purchased over an encrypted connection between mutually anonymous parties, paid for with difficult-to-trace virtual currency. Brokers and buyers in this exchange would face risks similar to what black marketeers do today, including undercover agents and dishonest dealings (perhaps made all the more likely due to the anonymous nature of these virtual-world transactions).

Some people will cheer for the end of control that connectivity and data-rich environments engender. They are the people who believe that information wants to be free,* and that greater transparency in all things will bring about a more just, safe and free world. For a time, WikiLeaks' cofounder Julian Assange was the world's most visible ambassador for this cause, but supporters of WikiLeaks and

---

* This dictum is commonly attributed to Stewart Brand, the founder and editor of the *Whole Earth Catalog,* recorded at the first Hackers' Conference, in 1984.

the values it champions come in all stripes, including right-wing libertarians, far-left liberals and apolitical technology enthusiasts. While they don't always agree on tactics, to them, data permanence is a failsafe for society. Despite some of the known negative consequences of this movement (threats to individual security, ruined reputations and diplomatic chaos), some free-information activists believe the absence of a delete button ultimately strengthens humanity's progress toward greater equality, productivity and self-determination. We believe, however, that this is a dangerous model, especially given that there is always going to be someone with bad judgment who releases information that will get people killed. This is why governments have systems and valuable regulations in place that, while imperfect, should continue to govern who gets to make the decision about what is classified and what is not.

We spoke with Assange in June 2011, while he was under house arrest in the United Kingdom. Our above-mentioned position aside, we must account for what free-information activists may try to do in the future, and therefore, Assange is a useful starting point. We will not revisit the ongoing debates of today (about which there are already many books and articles), which focus largely on the Western reaction to WikiLeaks, the contents of the cables that have been leaked, how destructive the leaks were and what punishments should await those involved in such activities. Instead, our interest is in the future and what the next phase of free-information movements—beginning with, but not restricted to, the Assange types—may try to achieve or destroy. Over the course of the interview, Assange shared his two basic arguments on this subject, which are related: First, our human civilization is built upon our complete intellectual record; thus the record should be as large as possible to shape our own time and inform future generations. Second, because different actors will always try to destroy or otherwise cover up parts of that shared history out of self-interest, it should be the goal of everyone who seeks and values truth to get as much as possible into the record, to prevent deletions from it, and then to make this record as accessible and searchable as possible for people everywhere.

Assange's is not a war on secrecy, per se—"There are all sorts of reasons why non-powerful organizations engage in secrecy," he told us, "and in my view it's legitimate; they need it because they're powerless"—but instead it is a fight against the secrecy that shields actions not in the public's interest. "Why are powerful organizations engaged in secrecy?" he asked rhetorically. The answer he offered is that the plans they have would be opposed if made public, so secrecy floats them to the implementation stage, at which point it's too late to alter the course effectively. Organizations whose plans won't incur public opposition don't carry that burden, so they don't need to be secretive, he added. As these two types of organizations battle, the one with genuine public support will eventually come out on top, Assange said. Releasing information, then, "is positive to those engaged in acts which the public supports and negative to those engaged in acts the public doesn't support."

As to the charge that those secretive organizations can simply take their operations off-line and avoid unwelcome disclosure, Assange is confident in his movement's ability to prevent this. Not a possibility, he said; serious organizations will always leave a paper trail. By definition, he explained, "systematic injustice is going to have to involve a lot of people." Not every participant will have full access to the plans, but each will have to know something in order to do his job. "If you take your information off paper, if you take it outside the electronic or physical paper trail, institutions decay," he said. "That's why all organizations have rigorous paper trails for the instructions from the leadership." Paper trails ensure that instructions are carried out properly; therefore, as Assange said, "if they internally balkanize so that information can't be leaked, there's a tremendous cost to the organizational efficiency of doing that." And inefficient organizations mean less powerful ones.

Openness, on the other hand, introduces new challenges for this movement of truth-seekers, from Assange's perspective. "When things become more open, then they start to become more complex, because people start hiding what they're doing—their bad behavior—through complexity," he said. He pointed to bureaucratic doublespeak and the offshore financial sector as clear examples. These systems are technically open, he said, but in fact are impenetrable; they are hard to attack but even harder to use efficiently. Obfuscation at this level, where the complexity is legal but still covering something up, is a much more difficult problem to solve than straightforward censorship.

Unfortunately, people like Assange and organizations like WikiLeaks will be well placed to take advantage of some of the changes in the next decade. And even supporters of their work are faced with difficult questions about the methods and implications of online disclosures, particularly as we look beyond the case study of WikiLeaks and into the future. One of the most difficult is the question of discretionary power: Who gets to decide what information is suitable for release, and what must be redacted, even temporarily? Why is it Julian Assange, specifically, who gets to decide what information is relevant to the public interest? And what happens if the person who makes such decisions is willing to accept indisputable harm to innocents as a consequence of his disclosures? Most people would agree that some level of supervision is necessary for any whistle-blowing platforms to serve a positive role in society, but there is no guarantee that supervision will be there (a glance at the recklessness of hackers* who publish others' personal information online in bulk confirms this).

If there is a central body facilitating the release of information, someone or some group of people, with their own ideas and biases, must be making those decisions. So long as humans, and not computers, are running things in our world, we will face these questions of judgment, no matter how transparent or technically sound the platforms are.

Looking ahead, some people might assume that the growth of connectivity around the world will spur a proliferation of WikiLeaks-like platforms. With more users and more classified or confidential information online, the argument goes, dozens of smaller secret-publishing platforms will emerge to meet the increase in supply and demand. A compelling and frightening idea, but wrong. There are natural barriers to growth in the field of whistle-blowing websites, including exogenous factors that limit the number of platforms that can successfully coexist. Regardless of what one thinks of WikiLeaks, consider all the things it needed in order to become a known, global brand: more than one geopolitically relevant large-scale leak to grab international attention; a track record of leaks to show commitment to the cause, to generate public trust and to give incentives to other potential leakers by demonstrating WikiLeaks' ability to protect them; a charismatic fig-

---

* While in the technical community the term "hacker" means a person who develops something quickly and with an air of spontaneity, we use it here in its colloquial meaning to imply unauthorized entry into systems.

urehead who could embody the organization and serve as its lightning rod, as Assange called himself; a constant upload of new leaks (often in bulk) to remain relevant in the public eye; and, not least, a broadly distributed and technically sophisticated digital platform for leakers, organization staff and the public to handle the leaked materials (while all remaining anonymous to one another) that could evade shutdown by authorities in multiple countries. It is very difficult to build such an intricate and responsive system, both technically and because the value of most components depends on the capabilities of others. (What good is a sophisticated platform without motivated leakers, or a set of valuable secrets without the system to discretely process and disseminate them?) The balance struck by WikiLeaks between public interest, private disclosure and technical protections took years to reach, so it is hard to imagine future upstarts, offshoots or rivals building an equivalent platform and brand much faster than they could—particularly now that authorities around the world are attuned to the threat such organizations pose.

Moreover, even if new organizations managed to build such platforms, it is highly unlikely that the world could support more than a handful at any given time. There are a few reasons for this. First, even the juiciest disclosures require a subsequent media cycle in order to have impact. If the landscape of secret-spilling websites became too decentralized, media outlets would find it difficult to keep track of these sites and their leaks, and to gauge their trustworthiness as sources. Second, leakers will naturally coalesce around organizations that they believe will generate maximum impact for their disclosures while providing them with the maximum amount of protection. These websites can compete for leakers, with promises of ever better publicity and anonymity, but it's only logical that a potential whistle-blower would look for successful examples and follow the lead of other leakers before him. What source would risk his chance, even his life, on an untested group? And organizations that cannot consistently attract high-level leaks will lose attention and funding, slowly but surely atrophying in the process. Assange described this dynamic from his organization's perspective as a positive one, providing a check on WikiLeaks as surely as it kept them in business. "Sources speak with their feet," he said. "We're disciplined by market forces."

Regionality may determine the future of whistle-blowing websites more than anything else. Governments and corporations in the West are, for the most part, now wise to the risks that lackluster cybersecurity allows, and though their systems are by no means impenetrable, significant resources are being invested in both the public and the private sector to better protect records, user data and infrastructure. The same is not true for most developing countries, and we can expect that as these populations come online in the next decade, some will experience their own version of the WikiLeaks phenomenon: sources with access to newly digitized records and the incentive to leak sensitive materials to cause a political impact. The ensuing storms may be limited to a particular country or region, but they will nonetheless be disruptive and significant for the environments they touch. They may even catalyze a physical revolution or riot. We should also expect the deployment of similar tactics from government authorities to combat such sites (even if the organizations and their servers are based elsewhere): filtering, direct attacks, financial blockades and legal prosecution.

Eventually, though, the technology used by these platforms will be so sophisticated that they will be effectively unblockable. When WikiLeaks lost its principal website URL, WikiLeaks.org, due to a series of distributed denial-of-service (DDoS) attacks and the pullout of its Internet service provider (which hosted the site) in 2010, its supporters immediately set up more than a thousand "mirror" sites (copies of the original site hosted at remote locations), with URLs like WikiLeaks.fi (in Finland), WikiLeaks.ca (in Canada) and WikiLeaks .info. (In a DDoS attack, a large number of compromised computer systems attack a single target, overloading the system with information requests and causing it to shut down, denying service to legitimate users.) Because WikiLeaks was designed as a distributed system—meaning its operations were distributed across many different computers, instead of concentrated in one centralized hub—shutting down the platform was much more difficult than it seemed to most laymen. Future whistle-blowing websites will surely move beyond mirror sites (copies of existing sites) and use new methods to replicate and obfuscate their operations to shield themselves from authorities. One way to accomplish this would be to create a storage system where fragments of files are copied and distributed in such a way that if one file directory is shut down, the files can be reassembled from those fragments. These platforms will develop new ways to ensure anonymous submission for potential leakers; WikiLeaks constantly updated its submission methods, warning users to avoid earlier cryptographic routes—among them SSL, or secure sockets layer, and hidden Tor service, using the highly encrypted Tor network—once they had determined that those were insufficiently secure.

And what of the individuals leading this charge? The Assanges of the world will still exist in the future, but their support bases will remain small. The more welcomed whistle-blowers of the future will be the ones who follow the example of people like Alexei Navalny, a Russian blogger and anticorruption activist, who enjoys much sympathy from many in the West. Disillusioned with Russia's liberal opposition parties, Navalny, a real-estate lawyer, started his own blog dedicated to exposing corruption in major Russian companies, initially supplying the disclosures himself by taking small stakes in the businesses and invoking shareholder rights to force them to share information. He later crowd-sourced his approach, instructing supporters to try to do the same, with some success. Eventually, his blog grew into a full-blown secret-spilling platform, where visitors were encouraged to donate toward its operating costs via PayPal. Navalny's profile grew as his collection of scoops swelled, most notably with a set of leaked documents that revealed the misuse of $4 billion at the state-owned oil pipeline company Transneft in 2010. By late 2011, Navalny's public stature placed him at the center of preelection protests, and his nickname for Vladimir Putin's United Russia party, the Party of Crooks and Thieves, had gone viral, adopted widely throughout the country.

Navalny's approach, at least in the beginning of his new activism, was distinctive in that for all his zeal he had not turned the focus of his whistle-blowing operation toward Putin himself. His targets had largely been commercial, although given that the Russian public and private sector are not always easily distinguished, the information implicated some government officials as well. Moreover, despite the harassment he experienced—he had been arrested, imprisoned, spied on and investigated for embezzlement—he remained free for years. His critics may have called him a liar, a hypocrite or a CIA stooge, but Navalny remained in Russia (unlike so many other high-profile Kremlin opponents) and his blog was not censored.

Some think Navalny did not constitute much of a threat to the

Kremlin; his name recognition among Russians remained quite low, though his supporters argue that such figures merely reflect low Internet penetration across the country and the success of state media censorship (Navalny was banned from appearing on state-run television). But a more interesting theory is that, for a time at least, Navalny found a way to toe the line as an anticorruption activist, knowing what to leak—and from whom—and what areas to avoid. Unlike prominent Putin critics, like the jailed billionaire Mikhail Khodorkovsky and the self-exiled oligarch Boris Berezovsky, Navalny seems to have found a way to challenge the Kremlin, while fighting corruption, without veering into overly sensitive areas that might place him in grave danger. (Short of a badly doctored photograph that appeared in a pro-Kremlin newspaper showing Navalny laughing with Berezovsky, there is little to suggest he has any ties to those critics.) His presence seemed to be tolerated by the Russian government until July 2012, when it deployed all available tools to discredit him, formally charging him with embezzlement in a case concerning a state-owned timber business in the Kirov region, where he had formerly worked as an advisor to the governor. The charges, carrying a maximum sentence of ten years in prison, reflected how much of a threat the resilient antigovernment protest movement had become. The world will continue to watch the trajectories of figures like Navalny to see whether his approach provides some measure of insulation from attack for digital activists.

There is also the frightening possibility that sites will emerge created by people who share the design and scale of these whistle-blower platforms but not their motivations. Rather than functioning as a clearinghouse for whistle-blowers, such platforms would serve as hosts to all manner of pilfered digital content—leaked active military operations, hacked bank accounts, stolen passwords and home addresses—without any particular agenda beyond anarchy. Operators of these sites would not be ideologues or political activists; they would be agents of chaos. Today, hackers and information criminals publish their ill-gotten gains fairly indiscriminately—the 150,000 Sony customer records released by the hacker group LulzSec in 2011 were simply made downloadable as a file through a peer-to-peer file-sharing service—but in the future, if a centralized platform emerged that offered them WikiLeaks-level security and publicity, it would present a real problem. Redaction, verification and other precautionary measures taken by WikiLeaks and its media partners would surely not be performed on these unregulated sites (indeed, Assange told us he redacted only to reduce the international pressure that was financially strangling him and said he would have preferred no redactions), and lack of judgment around sensitive materials might well get people killed. Information criminals would almost certainly traffic in bulk leaks in order to cause maximum disruption. To some extent, leaking selectively reflects purpose while releasing material in bulk is effectively thumbing one's nose at the entire system of secure information.

But context matters, too. How different would the reaction have been, from Western governments in particular, if WikiLeaks had published stolen classified documents from the regimes in Venezuela, North Korea and Iran? If Bradley Manning, the alleged source of WikiLeaks' materials about the United States government and military, had been a North Korean border guard or a defector from Iran's Revolutionary Guard Corps, how differently would politicians and pundits in the United States have viewed him? Were a string of whistle-blowing websites dedicated to exposing abuses within *those* countries to appear, surely the tone of the Western political class would shift. Taking into account the precedent President Barack Obama set in his first term in office—a clear "zero tolerance" approach toward unauthorized leaks of classified information from U.S. officials—we would expect that future Western governments would ultimately adopt a dissonant posture toward digital disclosures, encouraging them abroad in adversarial countries, but prosecuting them ferociously at home.

The Reporting Crisis

Where we get our information and what sources we trust will have a profound impact on our future identities. What's in store for the news in the Internet era is well-covered ground, and the battles we see today over monetization strategies and content syndication will continue to play out in the coming decade. But as technology lowers entry barriers in every industry, how will the media landscape as we know it today change?

It is manifestly clear that mainstream media outlets will increasingly find themselves a step behind in the reporting of news worldwide. These organizations simply cannot move quickly enough in a connected age, no matter how talented their reporters and stringers are, and how many sources they have. Instead, the world's breaking news will continually come from platforms like Twitter: open networks that facilitate information-sharing instantly, widely and in accessible packages. If everyone in the world has a data-enabled phone or access to one—a not-so-distant reality—then the ability to "break news" will be left to luck and chance, as one unwitting civilian in Abbottabad, Pakistan, discovered after he unknowingly live-tweeted the covert raid that killed Osama bin Laden.*

Eventually, this lag time—before the mainstream media can get the story—will alter the nature of audiences' loyalty, as readers and viewers seek more immediate methods of information delivery. Every future generation will be able to produce and consume more information than the previous one, and people will have little patience or use for media that cannot keep up. The loyalty that audiences retain will derive from the analysis and perspective these outlets offer, and, most critically, the trust they have in these institutions. These audiences will trust the credibility of the information, the accuracy of the analysis and the prioritization of news stories. In other words, some people will split their loyalty between new platforms for breaking news and established media organizations for the rest of the story.

News organizations will remain an important and integral part of society in a number of ways, but many outlets will not survive in their current form—and those that do survive will have adjusted their goals, methods and organizational structure to meet the changing demands of the new global public. As language barriers break down and cell towers rise, there will be no end to the number of new voices, potential sources, citizen journalists and amateur photographers looking to contribute. This is good: With so many news outlets scaling back their operations, particularly their international footprint, such outside contributors will be needed. The global audience benefits as well, through exposure to a greater range of issues and perspectives. The effect of having so many new actors involved, connected through a range of

---

* Among the tweets the Pakistani IT consultant Sohaib Athar sent the night of the bin Laden raid: "Helicopter hovering above Abbottabad at 1AM (is a rare event)."

online platforms into the great, diffuse media system, is that major media outlets will report less and validate more.

Reporting duties will become more widely distributed than they are today, which will expand the scope of coverage but probably reduce the quality on a net level. The role of the mainstream media will primarily become one of an aggregator, custodian and verifier, a credibility filter that sifts through all of this data and highlights what is and is not worth reading, understanding and trusting. Particularly for the elite—the business leaders, policymakers and intellectuals who rely on established media—validation will be critical, as will the media's ability to provide cogent analysis. In fact, the elite will probably rely *more* on established news organizations simply because of the massive swell of low-grade reporting and information in the system. Twitter can no more produce analysis than a monkey can type out a work of Shakespeare (although a heated Twitter exchange between two smart, credible people can come close); the strength of open, unregulated information-sharing platforms is their responsiveness, not their insight or depth.

Mainstream media outlets will have to find ways to integrate all of the new global voices they can now reach, a challenging but necessary task. Ideally, the business of journalism will become less extractive and more collaborative; in a story about rising tide levels in Bangkok, instead of just quoting a Thai river-cruise operator, the newspaper would link its article to the man's own news platform or personal live stream. Of course, the chance for error increases in the inclusion of new, untrained voices—many respected journalists today believe that a full-bodied embrace of citizen journalism is detrimental to the field, and their concerns are not unwarranted.

Global connectivity will introduce entirely new contributors to the supply chain. One new subcategory to emerge will be a network of local technical encryption specialists, who deal exclusively in encryption keys. Their value for journalists would not be content or source related but instead would provide the necessary confidentiality mechanisms between parties. Dissidents in repressive countries—for example, today's Belarus and Zimbabwe—will always be more willing to share their stories if they know they can do so safely and anonymously. Many people could potentially offer this technology, but local encryption specialists will be highly valued because trust is important. This is not too different from what we see throughout the Middle East today, where virtual private network (VPN) dealers roam busy marketplaces, along with other traders of illicit goods, to offer access to dissidents and rebellious youth to connect from their device to a secure network. Media organizations that cover international issues will rely on these scrappy young VPN and encryption dealers as they rely on foreign stringers to build their news coverage.

A new type of stringer will evolve as well. The conventional stringer today is an uncredited journalist whom newspapers pay to report, often from a foreign or unstable country. Stringers risk their lives to gain access to certain sources or visit dangerous places, taking these risks because professional reporters cannot or will not go there. An additional category of stringer may well emerge: men and women who deal exclusively in digital content and online sources. Instead of braving dangers on the ground, they'll take advantage of rising global connectivity to find, engage and extract information from sources they know only online. They would connect journalists with sources, as stringers do today. Obviously, given the additional layer of distance and obfuscation the virtual world presents, media outlets would have to exercise even greater caution than they usually do with regard to embellishment, validation of sources and ethics.

Imagine celebrities in the future starting their own news portal online about a particular ethnic conflict that they care deeply about. Perhaps they believe that the mainstream media isn't doing enough to publicize it or that it has gotten the narrative wrong. They decide to cut out the traditional middlemen and deliver stories directly to the public; let's call it Brangelina news. They hire their own people to work in the conflict zone, and they provide daily reports that their staff at home form into news articles to publish on their platform. Their overhead would be low, certainly lower than major news outlets, and they might not even need to compensate reporters and stringers, some of whom would work for free in exchange for the visibility. In short order, they become the ultimate source of information and news on the conflict because they both are highly visible and have built up enough credibility in their work that they can be taken seriously.

Mainstream media outlets will find such new serious competitors in the future—not just tweeters and amateur onsite observers—and that will complicate the media environment in this period. As we said, many will still favor and support the established news organizations, out of loyalty and trust in the institutions, and the serious work of journalism—the investigative reporting, the high-level interviews, the prescient contextualization of complicated events—will remain in the domain of the mainstream media. But for others, the diversification of content sources will represent a choice between a serious outlet and a "celebrity" outlet, and the seemingly insatiable appetite for tabloid-like content (in the United States, the U.K. and elsewhere) suggests that many consumers will probably choose the celebrity one. Visibility, not consistency or strength of content, will drive the popularity of such publishers.

Just as they do today with charities and business ventures, celebrities will look to starting their own media outlet as a logical extension of their "brand." (We are using as broad a definition of "celebrity" as possible here: We mean all highly visible public figures, which today could mean anyone from reality-TV stars to famous evangelical preachers.) To be sure, some of these new outlets will be solid attempts to contribute to public discourse, but many will be vapid and nearly content free, merely exercises in self-promotion and commercialized fame.

We will see a period in which people flock to these new celebrity outlets for their novelty value and to be part of a trend. Those that stay won't mind that the content and professionalism are a few notches below those of established media organizations. Media critics will decry these changes and lament the death of journalism, but this will be premature, because once the audience shifts, so too will the burden of reporting. If a celebrity outlet doesn't provide enough news, or consistently makes errors that are publicly exposed, the audience will leave. Loyalties are fickle when it comes to media, and this will only become truer as the field grows more crowded. If enough celebrity outlets lose the faith and trust of their audience, the resulting exodus will lead back to the professional media outlets, which will have undergone their own transformations (more aggregation, wider scope, faster response time) in the interim. Not all who left will return, just as not all who take issue with the mainstream media will jettison familiar information sources for new and trendy ones. Ultimately, it remains to be seen just how much impact these new celebrity competitors will have on the media landscape in the long term, but their emergence as players in the game of accruing viewers, readers and advertisers will undoubtedly cause a stir.

Expanded connectivity promises more than just challenges for media outlets; it offers new possibilities for the role of media more generally, particularly in countries where the press is not free. One reason that corrupt officials, powerful criminals and other malevolent forces in a society can continue to operate without fear of prosecution is that they control local information sources, either directly as owners and publishers or indirectly through harassment, bribery, intimidation or violence. This is as true in countries with largely state-owned media, like Russia, as it is in those where criminal syndicates hold enormous power and territory, like Mexico. The result—the lack of an independent press—reduces both accountability and the risk that public knowledge of misdeeds will lead to pressure and the political will to prosecute.

Connectivity can help upend such a power imbalance in a number of ways, and one of the most interesting ones concerns digital encryption and what it will enable underground or at-risk media organizations to do. Imagine an international NGO whose mission is to facilitate confidential reporting from places where it is difficult or dangerous to be a journalist. What differentiates this organization from others today, like watchdog groups and nonprofit media patrons, is the encrypted platform it builds and deploys to be used by media inside these countries. The platform's design is novel yet surprisingly simple. In order to protect the identities of journalists (who are the most exposed in the chain of reporting), every reporter for a given outlet is registered in the system with a unique code. Their names, mobile numbers and other identifiable details are encrypted behind this code, and the only people able to de-encrypt that information are key individuals at the NGO headquarters (not anyone at the news outlet), which, crucially, is based outside of the country. Inside the country, reporters are known only by this unique code—they use it to file stories and interact with their sources and local editors. As a result, if, for example, a journalist reports on an election irregularity in Venezuela (as many did during the October 2012 presidential election, although not anonymously), those charged with carrying out the president's dirty work have no way of knowing whom to target because they can't access the reporter's information, nor does anyone the reporter dealt with know who he or she really is. Media outlets don't maintain formal physical offices, since those could be targeted. Outlets necessarily have to vet their reporters initially, but after a journalist is introduced into the system, he is switched to a new editor (who has not met him) and his personal details evaporate into the platform.

The NGO outside of the country operates this platform from a safe distance, allowing the various participants to interact safely through a veil of encryption. Treating reporters in the same ways as confidential sources (protecting identities, preserving content) is not itself a new idea, but the ability to encrypt that identifiable data, and use an online platform to facilitate anonymous news-gathering, is only becoming possible now. The stories and other sensitive materials that journalists uncover can easily be stored in servers outside the country (someplace where there are strong legal protections around data), further limiting the exposure of those inside. Initially, perhaps this NGO would release its platform as a free product and operate it for different news outlets, financed by third-party donations. Eventually the NGO might take all of the working platforms and federate them, building a super-platform comprised of unidentifiable journalists from countries around the world. While we certainly do not advocate a popular shift toward anonymity, we assume in this case that the security situation is so dire and the society so repressive that the move is an act of desperation and necessity. An editor in New York would be able to log in, search for a reporter in Ukraine and find someone with a track record of published stories and even snippets from former colleagues. Without even knowing the journalist's name, the editor could rely on the available stories and the trust he has in this platform to decide whether to work with him. He could request an encrypted call with the reporter, also possible through the platform, to begin building a relationship.

This kind of disaggregated, mutually anonymous news-gathering system would not be difficult to build or maintain, and by encrypting the personal details of journalists (as well as their editors) and storing their reporting in remote servers, those who stand to lose as a more independent press emerges will become increasingly immobilized.

How does one retaliate against a digital platform, particularly in an age when everyone can read the news on their mobile devices? Connectivity is relatively low in many places that lack free media today, but as that changes, the reach of local reporting on sensitive matters will be even wider—international, in fact. These two trends—safer reporting backed by encryption and a wider readership due to gains in connectivity—ensure that even if a country's legal system is too corrupt or inept to properly prosecute bad actors, they can be publicly tried online through the media. Warlords operating in eastern Congo may not all be hauled into the International Criminal Court, but their lives will become more unpleasant if their every deed is captured and chronicled by unidentifiable and unreachable journalists, and the stories written about them travel to the far ends of the online world. At a minimum, other criminals who might otherwise do business with them will be deterred by their digital radioactivity, meaning they are too visible and under too much public scrutiny to be desirable business partners.

Privacy Revisited—Different Implications for Different Citizens

Security and privacy are a shared responsibility between companies, users and the institutions around us. Companies like Google, Apple, Amazon and Facebook are expected to safeguard data, prevent their systems from being hacked into and provide the most effective tools for users to maximize control of their privacy and security. But it is up to users to leverage these tools. Each day you choose not to utilize them you will experience some loss of privacy and security as the data keeps piling up. And you cannot assume there is a simple delete button. The option to "delete" data is largely an illusion—lost files, deleted e-mails and erased text messages can be recovered with minimal effort. Data is rarely erased on computers; operating systems tend to remove only a file's listing from the internal directory, keeping the file's contents in place until the space is needed for other things. (And even after a file has been overwritten, it's still occasionally possible to recover parts of the original content due to the magnetic properties of disc storage. This problem is known as "data remanence" by computer experts.) Cloud computing only reinforces the permanence of information, adding another layer of remote protection for users and their information.

Such mechanisms of retention were designed to save us from our own carelessness when operating computers. In the future, people will increasingly trust cloud storage—like ATMs in banks—over physical machinery, placing their faith in companies to store some of their most sensitive information, avoiding the risks of hard-drive crashes, computer theft or document loss. This multilayer backup system will make online interactions more efficient and productive, not to mention less emotionally fraught.

Near-permanent data storage will have a big impact on how citizens operate in virtual space. There will be a record of all activity and associations online, and everything added to the Internet will become part of a repository of permanent information. The possibility that one's personal content will be published and become known one day—either by mistake or through criminal interference—will always exist. People will be held responsible for their virtual associations, past and present, which raises the risk for nearly everyone since people's online networks tend to be larger and more diffuse than their physical ones. The good and bad behavior of those they know will affect them positively or negatively. (And no, stricter privacy settings on social-networking sites will not suffice.)

This will be the first generation of humans to have an indelible record. Colleagues of Richard Nixon may have been able to erase those eighteen and a half minutes of a tape recording regarding the Watergate break-in and cover up, but today's American president faces a permanent record of every e-mail sent from his BlackBerry, accessible to the public under the Presidential Records Act.

Since information wants to be free, don't write anything down you don't want read back to you in court or printed on the front page of a newspaper, as the saying goes. In the future this adage will broaden to include not just what you say and write, but the websites you visit, who you include in your online network, what you "like," and what others who are connected to you do, say and share.

People will become obsessively concerned about where personal information is stored. A wave of businesses and start-ups will emerge promising to offer solutions, from present-day applications such as

Snapchat, which automatically deletes a photo or message after ten seconds, to more creative solutions that also add a layer of encryption and a shorter countdown. At best, such solutions will only mitigate the risk of private information being released more broadly. Part of this is due to counter-innovations such as apps that will automatically take a screenshot of every message and photo sent faster than your brain can instruct your fingers to command your device. More scientifically, attempts to keep personal information private are always going to be defeated by attacking the analog hole, which stipulates that information must eventually be seen if it is to be consumed. As long as this holds true, there will always be the risk of someone taking a screenshot or proliferating the content.

If we are on the web we are publishing and we run the risk of becoming public figures—it's only a question of how many people are paying attention, and why. Individuals will still have some discretion over what they share from their devices, but it will be impossible to control what others capture and share. In February 2012, a young Saudi newspaper columnist named Hamza Kashgari posted an imaginary conversation with the Prophet Muhammad on his personal Twitter account, at one point writing that "I have loved aspects of you, hated others, and could not understand many more." His tweets sparked instant outrage (some people considered his posts blasphemous or a sign of apostasy, both serious sins in conservative Islam). He deleted them within six hours of posting—but not before thousands of angry responses, death threats and the creation of a Facebook group called "The Saudi People Demand Hamza Kashgari's Execution." Kashgari fled to Malaysia but was deported three days later to Saudi Arabia, where charges of blasphemy (a capital crime) awaited him. Despite his immediate apology after the incident and a subsequent August 2012 apology, the Saudi government refused to release him. In the future, it won't matter whether messages like these are public for six hours or six seconds; they will be preserved as soon as electronic ink hits digital paper. Kashgari's experience is just one of many sad and cautionary stories.

Data permanence will persist as an intractable challenge everywhere and for all people, as we said, but the type of political system and level of government control in place will greatly determine how it affects people. To examine these differences in detail, we'll consider an open democracy, a repressive autocracy and a failed state.

In an open democracy, where free expression and responsive governance feed the public's impulse to share, citizens will increasingly serve as judge and jury of their peers. More available data about everyone will only intensify the trends we see today: Every opinion will find space in an expansive virtual landscape, real-time updating will foster hyperactive social and civil spheres, and the ubiquity of social networking will allow everyone to play celebrity, paparazzo and voyeur, all at once. Each person will produce a voluminous amount of data about himself—his past and present, his likes and choices, his aspirations and daily habits. Like today, much of this will be "opt-in," meaning the user deliberately chooses to share content for some undefined social or commercial reason; but some of it won't be. Also like today, many online platforms will relay data back to companies and third parties about user activity without their express knowledge. People will share more than they're even aware of. For governments and companies, this thriving data set is a gift, enabling them to better respond to citizen and customer concerns, to precisely target specific demographics of the population, and, with the emergent field of predictive analytics, to predict what the future will hold.*

As we said earlier, never before will so much data be available to so many people. Citizens will draw conclusions about one another from accurate and inaccurate sources, from "legitimate" sources like LinkedIn profiles and "illegitimate" ones like errant YouTube comments long forgotten. More than a few aspiring politicians will fall on their swords as past behavior documented online is later brought to light. Certainly, with time, the normalization trend that softened public attitudes toward leaders' infidelity or past drug use—who can forget President Bill Clinton's caveat that he "didn't inhale"?—will take hold. Perhaps the voting public will shrug off a scandalous post or

---

* "Predictive analytics" is a young field of study at the intersection of statistics, data-mining and computer modeling. At its core, it uses data to make useful predictions about the future. For one example, predictive analytics could use data on ridership fluctuations on the New York City subway to predict how many trains would be needed on a given day, accounting for seasonality, employment and the weather forecast.

photo based on a time stamp that predates the candidate's eighteenth birthday. Public acceptance for youthful indiscretions documented on the Internet will move a few paces forward, but probably not until a painful liminal period passes. In some ways, this is the logical next stage of an era characterized by the loss of heroes. What began with mass media and Watergate will continue into the new digital age, where even more data about individuals, from nearly every part of their lives, is available for scrutiny. The fallibility of humans over a lifetime will provide an endless stream of details online to puncture mythical hero status.

Any would-be professional, particularly one in a position of trust, will have to account for his past if he is to get ahead. Would it matter to you if your family physician spent his weekends typing long screeds against immigrants, or if your son's soccer coach spent his twenties working as a tour guide in Bangkok's red-light district? This granular level of knowledge about our peers and leaders will produce unanticipated consequences within society. Documented pasts will affect many people in the workplace and in day-to-day life, and some citizens will spend their entire lives acutely aware of the potentially volatile parts of their lives, wondering what might surface online one day.

In democratic countries, corruption, crime and personal scandals will be more difficult to get away with in an age of comprehensive citizen engagement. The amount of information about people that enters the public domain—tax records, flight itineraries, phone geo-location sites (global-positioning-system data collected by a user's mobile phone) and so much more, including what is revealed through hacking—will undoubtedly provide countless suspicious citizens with more than enough to go on. Activists, watchdog groups and private individuals will work hand in hand to hold their leaders to account, and they'll have the tools necessary to determine whether what their government tells them is the truth. Public trust may initially fall, but it will emerge stronger as the next generation of leaders takes these developments into consideration.

When the scope of such changes becomes fully realized, large portions of the population will demand government action to protect personal privacy, at a much louder volume than anything we hear today. Laws will not change the permanence of digital information, but sensible regulations can install checks that will ensure some modicum of privacy for citizens who seek it. Today's government officials, with a few exceptions, don't understand the Internet—not its architecture or its manifold uses. This will change. In ten years, more politicians will understand how communication technologies work and how they empower citizens and other nongovernmental actors. The result will be public figures in government who can lead more informed debates on issues of privacy, security and user protection.

In democracies in the developing world, where both democratic institutions and technology are newer, government regulation around privacy will be more random. In each country, a particular incident will initially raise the issues at stake in dramatic fashion and drive public demand, similar to what has happened in the United States. A federal statute was passed in 1994 prohibiting state departments of motor vehicles from sharing personal information after a series of high-profile abuses of that information, including the murder of a prominent actress by a stalker. In 1988, following the leak of the late Judge Robert Bork's video-rental information during the Supreme Court nomination process, Congress passed the Video Privacy Protection Act, criminalizing disclosure of personally identifiable rental information without customer consent.*

While all of this digital chaos will be a nuisance to democratic societies, it will not destroy the democratic system. Institutions and polities will be left intact, if slightly battered. And once democracies determine the appropriate laws to regulate and control new trends, the result may even be an improvement, with a strengthened social contract and greater efficiency and transparency in society. But this will take time, because norms are not quick to change, and each democracy will move at its own pace.

Without question, the increased access to people's lives that the data revolution brings will give some repressive autocracies a dangerous advantage in targeting their citizens.

While this is a bad outcome and one we hope will be mitigated by

---

* Interestingly, the VPPA statute came into play in a Texas lawsuit in 2008, when a woman filed a class-action suit against Blockbuster for sharing her rental and sales record with Facebook without her permission. The parties settled.

developments discussed elsewhere in the book, we must understand that citizens living in autocracies will have to fight even harder for their privacy and security. Rest assured, demand for tools and software to help safeguard citizens living under digital repression will give rise to a growing and aggressive industry. And that is the power of this new information revolution: For every negative, there will be a counterresponse that has the potential to be a substantial positive. More people will fight for privacy and security than look to restrict it, even in the most repressive parts of the world.

But authoritarian regimes will put up a vicious fight. They will leverage the permanence of information and their control over mobile and Internet service providers to create an environment of heightened vulnerability for their citizens. What little privacy existed before will be long gone, because the handsets that citizens have with them at all times will double as the surveillance bugs regimes have long wished they could put in people's homes. Technological solutions will protect only a distinct technically savvy minority, and only temporarily.

Regimes will compromise devices before they are sold, giving them access to what everybody says, types and shares in public and in private. Citizens will be oblivious to how they might be vulnerable to giving up their own secrets. They will accidentally provide usable intelligence on themselves—particularly if they have an active online social life—and the state will use that to draw damning conclusions about who they are and what they might be up to. State-initiated malware and human error will give regimes more intelligence on their citizens than they could ever gather through non-digital means. Networks of citizens, offered desirable incentives by the state, will inform on their fellows. And the technology already exists for regimes to commandeer the cameras on laptops, virtually invade a dissident's home without his or her knowledge, and both listen to and watch everything that is said and done there.

Repressive governments will be able to determine who has censorship-circumvention applications on their handsets or in their homes, so even the non-dissident just trying to illegally download *The Sopranos* will come under increased scrutiny. States will be able to set up random checkpoints or raids to search people's devices for encryption and proxy software, the presence of which could earn them fines, jail time or a spot on a government database of offenders. Everyone who is known to have downloaded a circumvention measure will suddenly find life more difficult—they will not be able to get a loan, rent a car or make an online purchase without some form of harassment. Government agents could go classroom to classroom at every school and university in the country, expelling all students whose mobile-phone activity indicates that they've downloaded such software. Penalties could extend to these students' networks of family and friends, further discouraging that behavior for the wider population.

And, in the slightly less totalitarian autocracies, if the governments haven't already mandated "official" government-verified profiles, they'll certainly try to influence and control existing online identities with laws and monitoring techniques. They could pass laws that require social-networking profiles to contain certain personal information, like home address and mobile number, so that users are easier to monitor. They might build sophisticated computer algorithms that allow them to roam citizens' public profiles looking for omissions of mandated information or the presence of inappropriate content.

States are already engaging in this type of behavior, if somewhat covertly. As the Syrian uprising dragged on into 2013, a number of Syrian opposition members and foreign aid workers reported that their laptops were infected with computer viruses. (Many hadn't realized it until their online passwords suddenly stopped working.) Information technology (IT) specialists outside of Syria checked the discs and confirmed the presence of malware, in this case different types of Trojan horse viruses (programs that appear legitimate but are in fact malicious) that stole information and passwords, recorded keystrokes, took screenshots, downloaded new programs and remotely turned on webcams and microphones, and then sent all of that information back to an IP address which, according to the IT analysts, belonged to the state-owned telecom, Syrian Telecommunications Establishment. In this case, the spyware arrived through executable files (the user had to independently open a file to download the virus), but that doesn't mean the targeted individuals had been careless. One aid worker had downloaded a file, which appeared to be a dead link (meaning it no longer worked), in an online conversation with a person she thought was a verified opposition activist about the humanitarian need in the country. Only after the conversation did she learn to her chagrin that she had probably spoken with a government impersonator who possessed stolen or coerced passwords; the real activist was in prison.

People living under these conditions will be left to fend for themselves against the tag team of their government and its corrupt corporate allies. What governments can't build in-house, they can outsource to willing suppliers. Guilt by association will take on a new meaning with this level of monitoring. Just being in the background of a person's photo could matter if a government's facial-recognition software were to identify a known dissident in the picture. Being documented in the wrong place at the wrong time, whether by photo, voice or IP address, could land unwitting citizens in an unwanted spotlight. Though this scenario is profoundly unfair, we worry that it will happen all too often, and could encourage self-censoring behaviors among the rest of society.

If connectivity enhances the state's power, enabling it to mine its citizens' data with a fly-on-the-wall vantage point, it also constricts the state's ability to control the news cycle. Information blackouts, propaganda and "official" histories will fail to compete with the public's access to outside information, and cover-ups will backfire in the face of an informed and connected population. Citizens will be able to capture, share and remark upon an event before the government can decide what to say or do about it, and thanks to the ubiquity of cheap mobile devices, this grassroots power will be fairly evenly distributed throughout even large countries. In China, where the government has one of the world's most sophisticated and far-reaching censorship systems in place, attempts to cover up news stories deemed potentially damaging to the state have been missing the mark with increasing frequency.

In July 2011, the crash of a high-speed train in Wenzhou, in southeast China, resulted in the deaths of forty people and gave weight to a widely held fear that the country's infrastructure projects were moving too quickly for proper safety reviews. Yet the accident was downplayed by official channels, its coverage in the media actively minimized. It took tens of millions of posts on *weibos*, Chinese microblogs similar to Twitter, for the state to acknowledge that the crash had been the result of a design flaw and not bad weather or an electricity outage, as had previously been reported. Further, it was revealed that the government sent directives to the media shortly after the crash, specifically stating,

"There must be no seeking after the causes [of the accident], rather, statements from authoritative departments must be followed. No calling into doubt, no development [of further issues], no speculation and no dissemination [of such things] on personal microblogs!" The directives also instructed journalists to maintain a feel-good tone about the story: "From now on, the Wenzhou train accident should be reported along the theme of 'major love in the face of major disaster.'" But where the mainstream media fell in line, the microbloggers did not, leading to a deeply embarrassing incident for the Chinese government.

For a country like China, this mix of active citizens armed with technological devices and tight government control is exceptionally volatile. If state control relies on the perception of total command of events, every incident that undermines that perception—every misstep captured by camera phone, every lie debunked with outside information—plants seeds of doubt that encourage opposition and dissident elements in the population, and that could develop into widespread instability.

There may be only a handful of failed states in the world today, but they offer an intriguing model for how connectivity can operate in a power vacuum. Indeed, telecommunications seems to be just about the only industry that can thrive in a failed state. In Somalia, telecommunications companies have come to fill many of the gaps that decades of war and failed government have created, providing information, financial services and even electricity.

In the future, as the flood of inexpensive smart phones reaches users in failed states, citizens will find ways to do even more. Phones will help to enable the education, health care, security and commercial opportunities that the citizens' governments cannot provide. Mobile technology will also give much-needed intellectual, social and entertainment outlets for populations who have been psychologically traumatized by their environment. Connectivity alone cannot revert a failed state, but it can drastically improve the situation for its citizens. As we'll discuss later, new methods to help communities handle conflict and post-conflict challenges—developments like virtual institution building and skilled labor databases in the diaspora—will emerge to accelerate local recovery.

In power vacuums, though, opportunists take control, and in these cases connectivity will be an equally powerful weapon in their hands. Newly connected citizens in failed states will have all the vulnerabilities of undeletable data, but none of the security that could insulate them from those risks. Warlords, extortionists, pirates and criminals will—if they're smart enough—find ways to consolidate their own power at the expense of other people's data. This could mean targeting specific populations, such as wealthier subclans or influential religious leaders, with more precision and virtually no accountability. If the online data (say, transfer records for a mobile money platform) showed that a particular extended family received a comparatively large sum of money from relatives in the diaspora, local thugs could stop by and demand tribute—paid, probably, over a mobile money system as well. Today's warlords grow rich by acting as the requisite pass-through for all sorts of valuable resources, and in the future, while drugs, minerals and money will all still matter, so too will valuable personal data. Warlords of the future may not even use the data they have, instead selling it to outside parties willing to pay a premium. And, most important, these opportunists will be able to appear even more anonymous and elusive than they do today, because they'll unfortunately have the resources and incentive to get anonymity in ways ordinary people do not.

Power vacuums, warlords and collapsed states may sound like a foreign and unrelated world to many in Silicon Valley, but this will soon change. Today, technology companies constantly underscore their focus on, and responsibility to, the virtual world's version of citizenry. But as five billion new people come online, companies will find that the attributes of these users and their problems are much more complex than those of the first two billion. Many of the next five billion people live in impoverished, censored and unsafe conditions. As the providers of access, tools and platforms, technology companies will have to shoulder some of the physical world's burdens as they play out online if they want to stay true to the doctrine of responsibility to *all* users.

Technology companies will need to exceed the expectations of their customers in both privacy and security protections. It is unsurprising that the companies responsible for the architecture of the virtual world will shoulder much of the blame for the less welcome developments in our future. Some of the anger directed toward technology firms will be justified—after all, these businesses will be profiting from expanding their networks quickly—but much will be misplaced. It is, after all, much easier to blame a single product or company for a particularly evil application of technology than to acknowledge the limitations of personal responsibility. And of course there will always be some companies that allow their desire for profit to supersede their responsibility to users, though such companies will have a harder time achieving success in the future.

In truth, some technology companies are more acutely aware than others of the responsibility they bear toward their own users and the online community around the world; this is in part why nearly all online products and services today require users to accept terms and conditions and abide by those contractual guidelines. People have a responsibility as consumers and individuals to read a company's policies and positions on privacy and security before they willingly share information. As the proliferation of companies continues, citizens will have more options and thus due diligence will be more important than ever. A smart consumer will look not just at the quality of a product, but also at how easy that product makes it for you to control your privacy and security. Still, in the court of public opinion and environments where the rule of law is shaky, these preexisting stipulations count for little, and we can expect more attention to be focused on the makers and purveyors of such tools in the coming decades.

This trend will certainly affect how technology companies form, grow and navigate in what will certainly be a tumultuous period. Certain subsections of the technology industry that receive particularly negative attention will have trouble recruiting engineers or attracting users to and monetizing their products, despite the fact that such atrophying will not solve the problem (and will only hurt the community of users in the end, by denying them the full benefits of innovation). Thick skin will be a necessity for technology companies in the coming years of the digital age, because they will find themselves beset by public concerns over privacy, security and user protections. It simply won't be possible to avoid these discussions, nor will companies be able to avoid taking a position on the issues.

They'll also have to hire more lawyers. Litigation will always outpace genuine legal reform, as any of the technology giants fighting perpetual legal battles over intellectual property, patents, privacy and other issues would attest. Google encounters lawsuits from governments around the world with some frequency over alleged breaches of copyright or national laws, and it works hard to assure its users that Google serves their interests first and foremost, while staying within the boundaries of the laws itself. But if Google stopped all product development whenever it found itself faced with a government suit, it would never build anything.

Companies will have to learn how to manage public expectations of the possibilities and limits of their products. When formulating policies, technology companies will, like governments, increasingly have to factor in all sorts of domestic and international dynamics, such as the political risk environment, diplomatic relationships between states, and the rules that govern citizens' lives. The central truth of the technology industry—that technology is neutral but people are not—will periodically be lost amid all the noise. But our collective progress as citizens in the digital age will hinge on our not forgetting it.

Coping Strategies

People and institutions around the world will rise to meet the new challenges they face with innovative private- and public-sector coping strategies. We can loosely group them into four categories: corporate, legal, societal and personal.

Technology corporations will have to more than live up to their privacy and security responsibilities if they want to avoid unwanted government regulation that could stifle industry dynamism. Companies are already taking proactive steps, such as offering a digital "eject button" that allows users to liberate all of their data from a given platform; adding a preferences manager; and not selling personally identifying information to third parties or advertisers. But given today's widespread privacy and security concerns, there is still a great deal of work to be done. Perhaps a group of companies will make a pledge not to sell data to third parties, in a corporate treaty of sorts.

The second coping strategy will focus on the legal options. As the impact of the data revolution settles in, states will come under increasing pressure to protect their citizens from the permanence of what appears on the Internet and from their own newly exposed vulnerabilities. In democracies, this means new laws. They will be imperfect, overly idealistic and probably often quite rushed, but they will generally represent societies' best attempts to react effectively to the chaotic and unpredictable changes that connectivity produces.

As discussed above, the trail of information that will shape our online identities in the future begins well before any citizen has the judgment to understand it. The scrutiny that young people will face in the next decade will be unlike anything we've seen. If you think it is hard to get past a co-op board today, just imagine when it has the equivalent of your life story at hand. Because this development will affect a large portion of the population, there will be sufficient public pressure and political will to generate a range of new laws for the digital age.

As this next generation comes fully into adulthood, with digital documentation of every irresponsible thing they did during adolescence, it's hard to believe that some politicians won't champion the cause of sealing *virtual* juvenile records. Everything an individual shares before the age of eighteen might then become unusable, sealed and not for public disclosure on pain of fines or even prison. Laws would make it illegal for any employer, court, housing authority or university to take that content into account. Of course, these laws would be difficult to enforce, but their very presence would lend a hand in changing norms, so that most adolescent mishaps caught online may ultimately be viewed by society with the same lens as experimental drug and alcohol use.

Other laws may emerge as attempts to safeguard privacy and increase the liability for those releasing confidential information. Stealing someone's cell phone could be considered on a par with identity theft, and online intrusions (stolen passwords, hijacking accounts) could well carry the same charge as breaking and entering.* Each country will determine its own cultural threshold for what type of information is permissible to be shared, and what type is inappropriate or just too personal. What the Indian government considers obscene

---

* In the United States, the "trespass to chattels" tort has in some cases already been applied to cyberspace.

or perhaps pornographic, the French might let pass without a second thought. Consider the case of a society that is deeply concerned about privacy but is also saturated with camera-equipped smart phones and inexpensive camera drones that can be purchased at any toy store. The categories that exist for paparazzi photographers ("public" versus "private" space) could be extended and applied to everyone, with certain designated "safe zones" where photography requires a subject's consent (or, in the case of Saudi Arabia, consent from a female subject's male guardian). People would use specific apps on their phones to get permission, and because digital photos generate a time stamp and digital watermark, determining if someone took an illegal picture would be simple work. Digital watermarking refers to the insertion of bits into a digital image, audio or video file that contains copyright information about the file's owner—name, date, rights and so on. Watermarks act as protection against manipulation because, while they are invisible, they can be extracted and read with special software, so when tampering is suspected, technical experts can determine whether a file is indeed an unadulterated copy or not.

For the third type of coping strategy, at the societal level, we need to ask how non-state actors (such as communities and nonprofit organizations) will respond to the consequences of the data revolution. We think a wave of civil-society organizations will emerge in the next decade designed to shield connected citizens from their governments and from themselves. Powerful lobbying groups will advocate content and privacy laws. Rights organizations that document repressive surveillance tactics will call for better citizen protection. There will be support groups to help different demographics deal with the consequences of undeletable data. Educational organizations will try to reach school-age children to avoid over-sharing. ("Never give your data to a stranger.") The recent campaign in the United States against cyber-bullying is truly a harbinger of what is to come: broad public acknowledgment, grassroots social campaigns to promote awareness, and tepid political attempts to contain it. Within schools, we expect that teachers and administrators will treat cyber-bullying with the same weight and penalties as physical altercations, only instead of a child's being sent to the principal's office after recess, he will be sent there when he arrives in the morning for something he wrote online the previous night at home.

In addition to mitigating the negative consequences of a more connected world, non-state actors will be responsible for generating many of the most promising new ideas that harness these technological changes for the better. In developing countries, aid organizations are already leading the way with innovative pilot projects that capitalize on the growing global connectivity. During the 2011 famine in East Africa, the United States Agency for International Development (USAID) administrator Rajiv Shah reported that his organization was using a mix of mobile money platforms and the traditional *"hawala"* money-transfer system in Somalia to get past the violent Islamist group al-Shabaab's ban on aid for affected populations. (The *hawala* system is an Islamic-world network of trust-based money-transfer agents who operate outside of formal financial institutions.) The high rate of growth of mobile adoption and basic connectivity in the country has forged new opportunities for both the population and those seeking to help. Nonprofit and philanthropic organizations in particular will continue to push the boundaries of technology-driven solutions in the new digital age, well suited as they are to the task, being more flexible than government agencies and more able to absorb risk than businesses.

The fourth category of coping strategy is the personal. Citizens will demonstrate an increased reliance on anonymous peer-to-peer communication methods. In a world with no delete button, peer-to-peer (P2P) networking will become the default mode of operation for anyone looking to operate under or off the radar. Contemporary mobile P2P technologies like Bluetooth allow two physical devices to speak directly to each other rather than having to communicate over the Internet. This is in contrast to P2P file-sharing networks such as BitTorrent, which operate over the Internet. Common to both forms of peer-to-peer technologies is that users connect to each other (acting as both suppliers and receivers) without using a fixed third-party service. For citizens in the future, P2P networking will offer an enticing combination of instant communication and independence from third-party controls or monitoring.

All smart phones today are equipped with some form of peer-to-peer capability, and as the wave of cheap smart phones saturates the emerging markets in the next decade, even more people will be able to take advantage of these increasingly sophisticated tools. Bluetooth is already massively popular in many parts of the developing world because even very basic phones can often use it. In much of West Africa, where mobile adoption has vastly outpaced computer use and Internet growth, many people treat their phones like stereo systems because easy peer-to-peer sharing allows them to store, swap and listen to music entirely through their phones.

Mobile jukeboxes in Mali may be a response to specific infrastructure challenges, but people everywhere will begin to favor P2P networking, some for personal reasons (discomfort with undeletable records) and others for pragmatic ones (secure communications). Citizens in repressive societies already use common P2P communication platforms and encrypted messaging systems like Research in Motion (RIM)'s BlackBerry Messenger (BBM) to interact with less fear of government intrusion, and in the future, new forms of technologies that utilize P2P models will also become available to them.

Today, the discussions around wearable technologies are focused on a luxury market: wristwatches we'll wear that vibrate or apply a pulse when our alarm clock goes off (of which some versions already exist), earrings that monitor our blood pressure and so on.* New applications of augmented reality (AR) technology (the superimposing of touch, sound or images from the virtual world over a physical, real-world environment) promise even richer wearable experiences. In April 2012 Google unveiled its own AR prototype called Project Glass—eyeglasses with a built-in display over one eye that can convey information, handle messages through voice command and shoot and record video through its camera—and similar devices from other companies are on the way. In the future, the intersection of wearable technology, AR and peer-to-peer communications will combine sensory data, rich information channels and secure communications to generate exceptionally interesting and useful devices. In a country where religious police or undercover agents roam public areas, for example, good spatial awareness is critical, so a wearable-technology inventor will design

---

* Wearable technology overlaps with the similar emergent industry of haptic technology, but the two are not synonymous. Haptics refers to technology that interacts with a user's sense of touch, usually though pulses or the application of pressure. Wearable technologies often include many haptic elements but are not limited to them (like a jacket for cyclists that lights up in the evening); nor are all haptic technologies wearable.

a discreet wristwatch that its wearer can use to send a warning pulse to others around him when he spots a regime agent in his vicinity. An entirely new nonverbal language will emerge around sensory data—perhaps two pulses tell you a government agent is nearby, and three will mean "Run." Using GPS data, the watch would also share the location of its wearer with others, who might be wearing AR glasses that could identify which direction the agent is coming from. All these communications will be peer-to-peer. This makes them more secure and reliable than technologies that depend on being connected to the Internet.

Your device will know things about your surroundings that you have no way of knowing on your own: where people are, who they are and what their virtual profiles contain. Today, users already share their iTunes libraries with strangers over Wi-Fi networks, and in the future, they'll be able to share much more. In places like Yemen, where socially conservative norms limit many teenagers' ability to socialize with the opposite sex, young people may elect to hide their personal information on peer-to-peer networks when at home or at the mosque—who knows who could be looking?—but reveal it when in public parks and cafés, and at parties.

Yet P2P technology is a limited replacement for the richness and convenience of the Internet, despite its myriad advantages. We often need stored and searchable records of our activities and communications, particularly if we want to share something or refer to it later. And, unfortunately, not even P2P communications are a perfect shield against infiltration and monitoring. If authorities (or criminal organizations) can identify one side of a conversation they can usually find the other party as well. This is true for messaging, voice-over-Internet-protocol (VoIP) calls—meaning phone calls over the Internet (e.g., Google Voice and Skype) and video chats. Users assume they are safe, but unless the exchange is encrypted, anyone with access to intermediate parts of the network can listen in. For instance, the owner of a Wi-Fi hot spot can listen to any unencrypted conversations of users connected to the hot spot. One of the most insidious forms of cyber attack that P2P users can encounter is known as a "man-in-the-middle" attack, a form of active eavesdropping. In this situation a third-party attacker inserts himself between two participants in a conversation and automatically relays messages between them, without either participant realizing it. This third party acts like an invisible intermediary, having tricked each participant into believing that the attacker is actually the other party of the conversation. So as the conversation occurs (whether through text, voice or video), that third-party attacker can sit back and watch, occasionally siphoning off information and storing it elsewhere. (Or, more maliciously, the attacker could insert false information into the conversation.) Man-in-the-middle attacks occur in all protocols, not just peer-to-peer, yet they seem all the more malicious in P2P communications simply because people using those platforms *believe* they are secure.

And even the protection that encryption offers isn't a sure bet, especially given some of the checks that will still exist in the physical realm. In the United States, the FBI and some lawmakers have already hinted at introducing bills that would force communications services like BlackBerry and Skype to comply with wiretap orders from law-enforcement officials, introducing message-interception capabilities or providing keys that enable authorities to unscramble encrypted messages.

P2P networking has a history of challenging governments, especially around copyright issues for democracies (e.g., Napster, Pirate Bay) and political dissent for autocracies (e.g., Tor). In the United States, the pioneer of P2P file sharing, Napster, was shut down in 2001 by an injunction demanding that the company prevent all trading of copyrighted material on its network. (Napster told a district court that it was capable of blocking the transfer of 99.4 percent of copyrighted material, but the court said that wasn't good enough.) In Saudi Arabia and Iran, religious police have found it extremely difficult to prevent young people from using Bluetooth-enabled phones to call and text complete strangers within range, oftentimes for the purpose of flirting, but also for close-proximity coordination between protesters. Unless all mobile devices in the country are confiscated (a task the secret police realize is impossible), the flirtatious Saudi and Iranian youth have at least one small edge on their state-sponsored babysitters.

BlackBerry mobile devices offer both encrypted communication and telephone services, and the unique encryption they offer users has led many governments to target them directly. In 2009, the United Arab Emirates' partially state-owned telecom Etisalat sent nearly 150,000 of its BlackBerry users a prompt for a required update for "service enhancements." These enhancements were actually spyware that allowed unauthorized access to private information stored on users' phones. (When this became public knowledge, the maker of BlackBerry, RIM, distanced itself from Etisalat and told users how to remove the software.) Just a year later, the U.A.E. and its neighbor Saudi Arabia both called for bans on BlackBerry phones altogether, citing the country's encryption protocol. India chimed in as well, giving RIM an ultimatum to provide access to encrypted communications or see its services suspended. (In all three countries, the ban was averted.)

Repressive states will display little hesitation in their attempts to ban or gain control of P2P communications. Democratic states will have to act more deliberately. We already have a prominent example of this in the August 2011 riots in the United Kingdom. British protesters rallied to demand justice for twenty-nine-year-old Mark Duggan, who had been shot and killed by British police in Tottenham. Several days later the crowds turned violent, setting fire to local shops, police cars and a bus. Violence and looting spread across the country over subsequent nights, eventually reaching Birmingham, Bristol and other cities. The riots resulted in five deaths, an estimated £300 million ($475 million) in property damage and a great deal of public confusion. The scale of the disorder across the country—as well as the speed with which it spread—caught the police and government wholly off guard, and communication tools like Twitter, Facebook and particularly Black-Berry were singled out as a major operational factor in the spread of the riots. While the riots were occurring, the MP for Tottenham called on BlackBerry to suspend its messaging service during night hours to stop the rioters from communicating. When the violence had subsided, the British prime minister, David Cameron, told Parliament he was considering blocking these services altogether in certain situations, particularly "when we know [people] are plotting violence, disorder and criminality." His goal, he said, was to "give the police the technology to trace people on Twitter or BBM, or close it down." (After meeting with industry representatives, Cameron said industry cooperation with law enforcement was sufficient.)

The examples of the U.A.E. and the U.K. illustrate real concern on the part of governments, but it is important to clarify that this concern has been about encryption and social networking. In the future, however, communication will also take place on mobile P2P networks, meaning that citizens will be able to network without having to rely on the Internet (this was not the case in the U.A.E. and the U.K.). It stands to reason that every state, from the least democratic to the most, may fight the growth of device-to-device communication. Governments will claim that without restrictions or loopholes for special circumstances, capturing criminals and terrorists (among other legitimate police activities) and prosecuting them will become more difficult, planning and executing crimes will be easier and a person's ability to publish slanderous, false or other harmful information in the public sphere without accountability will improve. Democratic governments will fear uncontrollable libel and leaking, autocracies internal dissent. But if illegal activity is the primary concern for governments, the real challenge will be the combination of virtual currency with anonymous networks that hide the physical location of services. For example, criminals are already selling illegal drugs on the Tor network in exchange for Bitcoins (a virtual currency), avoiding cash and banks altogether. Copyright infringers will use the same networks.

As we think about how to address these kinds of challenges, we cannot afford to take a black-and-white view; context matters. For example, in Mexico, drug cartels are among some of the most effective users of anonymous encryption, both P2P and through the Internet. In 2011, we met with Bruno Ferrari, then the country's secretary of the economy, and he described to us how the Mexican government has struggled to engage the population in the fight against the cartels— fear of retribution is enough to prevent people from reporting crimes or tipping off law enforcement to cartel activity in their neighborhoods. Corruption and untrustworthiness in the police department further limit the options for citizens. "Without anonymity," Ferrari told us, "there is no clear mechanism in which people can trust the police and report the crimes committed by the drug cartels. True anonymity is vital to getting the citizens to be part of the solution." The drug cartels were already using anonymous communications, so anonymity levels the playing field. "The arguments behind restricting anonymous encryption make sense," he added, "but just not in Mexico."

Police State 2.0

All things considered, the balance of power between citizens and their governments will depend on how much surveillance equipment a government is able to buy, sustain and operate. Genuinely democratic states may struggle to deal with the loss of privacy and control that the data revolution enables, but as a result they will have more empowered citizens, better politicians and stronger social contracts. Unfortunately, the majority of states in the world are either not democratic or democratic in name only, and the relative impact of connectivity—both positive and negative—for citizens in those countries will be far greater than we'll see elsewhere.

In the long run, the presence of communication technologies will chip away at most autocratic governments, since, as we have seen, the odds against a restrictive, information-shy regime dealing with an empowered citizenry armed with personal fact-checking devices get progressively worse with each embarrassing incident. In other words, it's no coincidence that today's autocracies are for the most part among the least connected societies in the world. In the near term, however, such regimes will be able to exploit the growth of connectivity to their advantage, as they already exploit the law and the media. There is a trend in authoritarian governance to harness the power of connectivity and data, rather than ban information technology out of fear, a shift from totalitarian obviousness to more subtle forms of control that the journalist William J. Dobson captured in his excellent book *The Dictator's Learning Curve.* As Dobson describes it, "Today's dictators and authoritarians are far more sophisticated, savvy, and nimble than they once were. Faced with growing pressures, the smartest among them neither hardened their regimes into police states nor closed themselves off from the world; instead, they learned and adapted. For dozens of authoritarian regimes, the challenge posed by democracy's advance led to experimentation, creativity and cunning." Dobson identifies numerous avenues through which modern dictators consolidate power while feigning legitimacy: a quasi-independent judicial system, the semblance of a popularly elected parliament, broadly written laws that are applied selectively and a media landscape that allows for an opposition press as long as regime opponents understand where the unspoken limits are. Unlike the strongman regimes and pariah states of old,

Dobson writes, modern authoritarian states are "conscious, man-made projects that must be carefully built, polished, and reinforced."

But Dobson covers only a small number of case studies in his work and we are less certain that the new digital age will yield such advantages to *all* autocratic regimes. How dictators handle connectivity will greatly determine their future in the new digital age, particularly if their states want to compete for status and business on the global stage. The centralization of power, the delicate balancing of patronage and repression, the outward projection of the state itself—every element of autocratic governance will depend on the control that regimes have over the virtual world their population inhabits.

In the span of a decade, the world's autocracies will go from having a minority to a majority of their citizens online, and for dictators looking to stay in power, this will be a turbulent transition. Yet building the kind of system that can monitor and contain all types of dissident energy is thankfully not easy and will require very specialized solutions, expensive consultants, technologies not widely available and a great deal of money. Cell towers, servers and microphones will be needed, as well as large data centers to store information; specialized software will be necessary to process the data gathered; trained people will have to operate all of this, and basic resources like electricity and connectivity will need to be constantly and abundantly available. If autocrats want to build a surveillance state, it's going to cost them— we hope more than they can afford.

There are some autocracies with poor populations but vast amounts of oil, minerals or other resources that they can trade. As in the arms-for-minerals trade, we can imagine the growth of a technology-for-minerals exchange between technology-poor but resource-rich countries (Equatorial Guinea is one example) and technology-rich but resource-hungry countries (China is an obvious one). Not many states will be able to pull off this kind of trade, and hopefully those that do will not be able to sustain or effectively operate what they have.

Once the infrastructure is in place, repressive regimes will need to manage the glut of information they acquire with the help of supercomputers. In countries where connectivity was established early, governments have had time to acclimate to the types of data their citizens produce; the pace of technological adoption and progress has been somewhat gradual. But these newly wired regimes will not have that luxury; they'll need to move quickly to make use of their data if they want to be effective in its management. To address this, they'll build powerful computer banks with much faster processing power than the average laptop, and they'll buy or build software that facilitates the data-mining and real-time monitoring they desire. Everything a regime would need to build an incredibly intimidating digital police state is commercially available now, and export restrictions are currently insufficiently monitored and enforced.

Once one regime builds its surveillance state, it will share what it learned with others. We know that autocratic governments share information, governance strategies and military hardware, and it's only logical that the configuration that one state designs will (if it works) proliferate among its allies and assorted others. Companies that sell data-mining software, surveillance cameras and other products will flaunt their work with governments to attract new business.

The most important form of data to collect for an autocrat isn't Facebook posts or Twitter comments—it's biometric information. "Biometric" refers to information that can be used to uniquely identify individuals through their physical and biological attributes. Fingerprinting, photographs and DNA testing are all familiar biometric data types today. Indeed, the next time you visit Singapore, you might be surprised to find that airport security requires both a filled-out customs form *and* a scan of your voice. In the future, voice-recognition and facial-recognition software will largely surpass all of these earlier forms in accuracy and use.

The facial-recognition systems of today use a camera to zoom in on an individual's eyes, mouth and nose, and extract a "feature vector," which is a set of numbers that describes key aspects of the image, such as the precise distance between the eyes. (Remember, in the end, digital images are just numbers.) Those numbers can be fed back into a large database of faces in search of a match. To many this sounds like science fiction, and it's true that the accuracy of this software is limited today (by, among other things, pictures shot in profile), but the progress in this field in just the past few years is remarkable. A team at Carnegie Mellon demonstrated in a 2011 study that the combination of "off-the-shelf" facial-recognition software and publicly available online data can match a large number of faces very quickly, thanks to technical advancements like cloud computing. In one experiment, unidentified pictures from dating sites (where people often use pseudonyms) were compared with profile shots from social-networking sites, which can be publicly accessed on search engines (i.e., no log-in required), yielding a statistically significant result. It was noted in the study that it would be unfeasible for a human to do this search manually, but with cloud computing, it takes just seconds to compare millions of faces. The accuracy improves regarding people with many pictures of themselves available online—which, in the age of Facebook, is practically everyone.

Like so many technological advances, the promise of comprehensive biometric data offers innovative solutions to entrenched sociopolitical problems—and it makes dictators salivate. For each repressive regime that gathers biometric data to better oppress its population, however, a similar investment will be made by an open, stable and progressive country for very different reasons.

India's unique identification (UID) program is the largest biometric identification undertaking in the world. Constituted in 2009, the campaign, collectively called Aadhaar (meaning "foundation" or "support"), aims to provide every Indian citizen—1.2 billion and counting—with a card that includes a unique twelve-digit identity and an embedded computer chip that contains a person's biometric data, including fingerprints and iris scans. This vast program was conceived as a way to solve the problems of inefficiency, corruption and fraud endemic in the existing system, in which overlapping jurisdictions resulted in up to twenty different forms of identification issued by various local and national agencies.

Many in India believe that as the program progresses, Aadhaar will help citizens who have been excluded from government institutions and aid networks. For castes and tribes traditionally lowest on the socioeconomic scale, Aadhaar represents a chance to receive state aid like public housing and food rations—things that had been technically available but still out of reach, since many potential recipients lacked identification. Others who had trouble obtaining identification, like internal migrant workers, will be able to open a bank account, obtain a driver's license, apply for government support, vote and pay taxes with Aadhaar. When enrolling in the scheme, an individual may open a bank account that is tied to his or her UID number. This enables the government to easily track subsidies and benefits.

In a political system racked by political corruption and crippled by its own sheer size—less than 3 percent of the Indian population is registered to pay income tax—this effort seems like a possible win-win for all honest parties. Poor and rural citizens gain an identity, government systems become more efficient and all aspects of civic life (including voting and paying taxes) become more transparent and inclusive. But Aadhaar has its detractors, people who consider the program Orwellian in scope and character and a ploy to enhance the surveillance capacities of the Indian state at the expense of individual freedoms and privacy. (Indeed, the government can use Aadhaar to track the movements, phones and monetary transactions of suspected terrorists.) These detractors also point out that Indians do not have to have an Aadhaar card, since public agencies aren't allowed to require one before providing services. Concerns over whether the Indian government is intruding on civil liberties echo those of opponents of a similar project in the United Kingdom, the Identity Cards Act of 2006. (After a several-year struggle to implement the program, Britain's newly elected coalition government scrapped the plan in 2010.)

In India, these concerns seem to be outweighed by the promise of the plan's benefits, but their presence in the debate proves that even in a democracy, public apprehension over the impact of large biometric databases, and whether they'll ultimately serve the citizens or the state, exists. So what happens when less democratic governments begin collecting biometric data in earnest? Many already have, beginning with passports.

States won't be the only ones trying to acquire biometric data. Warlords, drug cartels and terrorist groups will seek to build or access biometric databases in order to track recruits, monitor potential victims and keep an eye on their own organizations. The same logic applies here as to dictators: If they have something to trade, they can get the technology.

Given the strategic value of these databases, states will need to prioritize protection of their citizens' information just as they would safeguard weapons of mass destruction. Mexico is currently moving toward a biometric data system for its population in order to improve its law-enforcement functionality, better monitor its borders and identify criminals and drug-cartel leaders. But since the cartels have already infiltrated large swaths of the police and national institutions, there is a very real fear that somehow an unauthorized actor could gain access to the valuable biometric data of the Mexican population. Eventually, some illicit group will successfully steal or illegally acquire a biometric database from a government, and maybe only when that happens will states fully invest in high-level security measures to protect this data.

All societies will reach agreement on the need to keep biometric data out of the hands of certain groups, and most will try hard to keep individual citizens from gaining access as well. Regulation will, like regulation of other types of user data, vary by country. In the European Union, which already boasts a series of robust biometric databases, member states are required by law to ensure that no individual's right to privacy is violated. States must get the full and informed consent of citizens before they can enter biometric information into the system, leaving citizens the option to revoke consent in the future without penalty. Member states are further required to hear complaints and see that victims are compensated. The United States will probably adopt similar laws due to shared privacy concerns, but in repressive countries, it's likely that such databases will be controlled by the ministry of the interior, ensuring that they are primarily used as a tool for the police and security forces. Government officials in those regimes will also have access to facial-recognition software, databanks of citizens' personal information and real-time surveillance methods through people's technological devices. Secret police will often find a handset more valuable than a gun.

For all of the discussions about privacy and security, we rarely look at the two together and ask the question What makes people nervous about the Internet? From the world's most repressive societies to those that are the most democratic, citizens are nervous about the unknowns, the dangers and crises that come with entangling their lives in a web of connected strangers. For those who are already connected, living in both the physical and the virtual worlds has become part of who we are and what we do. As we grow accustomed to this change, we also learn that the two worlds are not mutually exclusive, and what happens in one has consequences in the other.

What seem like defined debates today over security and privacy will broaden to questions of who controls and influences virtual identities and thus citizens themselves. Democracies will become more influenced by the wisdom of crowds (for better or for worse), poor autocracies will struggle to acquire the necessary resources to effectively extend control into the virtual world, and wealthier dictatorships will build modern police states that tighten their grip on citizens' lives. These changes will spur new behaviors and progressive laws, but given the sophistication of the technologies involved, in most cases citizens stand to lose many of the protections they feel and rely upon today. How populations, private industry and states handle the forthcoming changes will be highly determined by their social norms, legal frameworks and particular national characteristics.

We will now turn to a discussion of how global connectivity will affect the way states operate, negotiate and wrestle with each other. Diplomacy has never been as interesting as it will be in the new digital age. States, which are constantly playing power politics in the international system, will find themselves having to retool their domestic and foreign policies in a world where their physical and virtual tactics are not always aligned.

CHAPTER 3

The Future of States

What do we talk about when we talk about the Internet? Most people have only a vague sense of how the Internet works, and in most cases that's fine. The majority of users don't need to understand its internal architecture or how a hash function works in order to interface fluidly with the online world. But as we turn to a discussion about how state power affects, and is affected by, the Internet, some basic knowledge will help make clear a few of the more conceptually difficult scenarios that come into play.

As it was initially conceived, the Internet is a network of networks, a huge and decentralized web of computer systems designed to transmit information using specific standard protocols. What the average user sees—websites and applications, for example—is really the flora and fauna of the Internet. Underneath, millions of machines are sending, processing and receiving data packets at incredible speed over fiber-optic and copper cables. Everything we encounter online and everything we produce is ultimately a series of numbers, packaged together, sent through a series of routers located around the world, then reassembled at the other end.

We have often described the Internet as a "lawless" space, ungoverned and ungovernable by design. Its decentralized makeup and constantly mutating interlinking structure make government attempts to "control" it futile. But states have an enormous amount of power over the *mechanics* of the Internet in their own countries. Because states have power over the physical infrastructure connectivity requires—the transmission towers, the routers, the switches—they control the entry, exit and waypoints for Internet data. They can limit content, control what hardware people are allowed to use and even create separate Internets. States and citizens both gain power from connectivity, but not in the same manner. Empowerment for people comes from what they have access to, while states can derive power from their position as gatekeeper.

So far we have focused mostly on what will happen when billions more people come online—How will they use the Internet? What kinds of devices will they use? How will their lives change?—but we haven't yet said what their Internet will look like, or how states will make the most of it in their own physical and virtual dealings with other states and with their own people. This will increasingly matter, as populations with different alphabets, interests and sets of norms become connected, and as their governments bring their own interests, grudges and resources to the table. Perhaps the most important question in ten years' time won't be if a society uses the Internet, but which version of it they use.

As more states adapt to having large portions of their populations online, they'll strive to maintain control, both internally and on the world stage. Some states will emerge stronger—more secure and with greater influence—from this transition into the virtual age, benefiting from strong alliances and smart uses of digital power, while others will struggle just to keep up with and adapt to technological changes both domestically and internationally. Friendships, alliances and enmities between states will extend into the virtual world, adding a new and intriguing dimension to traditional statecraft. In many ways, the Internet could ultimately be seen as the realization of the classic international-relations theory of an anarchic, leaderless world. Here's how we think states will respond to each other and to their citizens.

The Balkanization of the Internet

As we said, every state and society in the world has its own laws, cultural norms and accepted behaviors. As billions of people come online in the next decade, many will discover a newfound independence—in ideas, speech and conversation—that will test these boundaries. Their governments, by contrast, would largely prefer that these users encounter a virtual world that allows the powers that be to mirror their physical control, an understandable if fundamentally naïve notion. Each state will attempt to regulate the Internet, and shape it in its own image. The impulse to project laws from the physical world into the virtual one is universal among states, from the most democratic to the most authoritarian. What states can't build in reality they'll try to fashion in virtual space, excluding those elements of society that they dislike, the content that contravenes laws and any potential threats they see.

The majority of the world's Internet users encounter some form of censorship—also known by the euphemism "filtering"—but what that actually looks like depends on a country's policies and its technological infrastructure. Not all or even most of that filtering is political censorship; progressive countries routinely block a modest number of sites, such as those featuring child pornography.

In some countries, there are several entry points for Internet connectivity, and a handful of private telecommunications companies control them (with some regulation). In others, there is only one entry point, a nationalized Internet service provider (ISP), through which all traffic flows. Filtering is relatively easy in the latter case, and more difficult in the former. Differences in infrastructure like these, combined with cultural particularities and objectives of filtering, account for the patchwork of systems around the world today.

In most countries, filtering is conducted at the ISP level. Typically, governments put restrictions on the gateway routers that connect the country and on DNS (domain name system) servers. This allows them to either block a website altogether (e.g., YouTube in Iran) or process web content through "deep-packet inspection." With deep-packet inspection, special software allows the router to look inside the packets of data that pass through it and check for forbidden words, among other things (the use of sentiment-analysis software to screen out negative statements about politicians, for example), which it can then block. Neither technique is foolproof; users can access blocked sites with circumvention technologies like proxy servers (which trick the routers) or by using secure https encryption protocols (which enable private Internet communication that, at least in theory, cannot be read by anyone other than your computer and the website you are accessing), and deep-packet inspection rarely catches every instance of banned content. The most sophisticated censorship states invest a great deal of resources to build these systems, and then heavily penalize anyone who tries to get around them.

When technologists began to notice states regulating and projecting influence online, some warned against a "balkanization of the Internet," whereby national filtering and other restrictions would transform what was once the *global* Internet into a connected series of nation-state networks.* The World Wide Web would fracture and fragment, and soon there would be a "Russian Internet" and an "American Internet" and so on, all coexisting and sometimes overlapping but, in important ways, separate. Each state's Internet would take on its national characteristics. Information would largely flow within countries but not across them, due to filtering, language or even just user preference. (Evidence shows that most users tend to stay within their own cultural spheres when online, less for reasons of censorship than because of shared language, common interest and convenience. The online experience can also be faster, as network caching, or temporarily storing content in a local data center, can greatly increase the access speed for users.) The process would at first be barely perceptible to users, but it would fossilize over time and ultimately remake the Internet.†

The first stage of this process, aggressive and distinctive filtering, is under way. It's very likely that some version of the above scenario will occur, but the degree to which it does will greatly be determined by what happens in the next decade with newly connected states—which path they choose, whom they emulate and work together with, and what their guiding principles turn out to be. To expand on these variations, let's look at a few different approaches to filtering in today's world. We've identified at least three models: the blatant, the sheepish, and the politically and culturally acceptable.

First, the blatant: China is the world's most active and enthusiastic

---

* We recommend the 2006 book *Who Controls the Internet?: Illusions of a Borderless World,* by Jack Goldsmith and Tim Wu, which puts forth this scenario with great clarity.

† Internet Balkans, as we refer to them, are different than intranets. An intranet uses the same Internet protocol technology but is limited to a network within an organization or local area, instead of a network of other networks. Corporate intranets are often protected from unauthorized external access by firewalls or other gateway mechanisms.

filterer of information. Entire platforms that are hugely popular else-where in the world—Facebook, Tumblr, Twitter—are blocked by the Chinese government. Particular terms like "Falun Gong"—the name of the banned spiritual group in China associated with one flank of the opposition—are simply absent from the country's virtual public space, victims of official censorship or widespread self-censorship. On the Chinese Internet, you would be unable to find information about politically sensitive topics like the Tiananmen Square protests, embarrassing information about the Chinese political leadership, the Tibetan rights movement and the Dalai Lama, or content related to human rights, political reform or sovereignty issues. When it comes to these topics, even some of the best-known Western media outlets fall victim to censorship. Bloomberg News was blocked in both En-glish and Chinese following its June 2012 exposé on the vast family fortune of the then vice-president (and now president), Xi Jinping. Four months later, *The New York Times* experienced a similar fate after publishing a similar story about the then premier, Wen Jiabao. Unsurprisingly, information about censorship circumvention tools is also blocked. We learned how comprehensive and particular Chinese censorship authorities could be when, following a contentious trip by Google's executive chairman, Eric, to Beijing in 2011, all traces of his visit were wiped from the Chinese Internet, while media coverage of his trip remained accessible everywhere else.

To the average Chinese user, this censorship is seamless—without prior knowledge of events or ideas, it would appear that they never existed. Further complicating matters, the Chinese government is not above taking a more proactive approach to online content: one esti-mate in 2010 suggested that Chinese officials had hired nearly three hundred thousand "online commenters" to write posts praising their bosses, the government and the Communist Party. (This kind of activ-ity is often called Astroturfing—i.e., fake grassroots participation—and is a popular tactic with public-relations firms, advertising agencies and election campaigns around the world.)

China's leadership doesn't hesitate to defend its strict censorship policies. In a white paper released in 2010, the government calls the Internet "a crystallization of human wisdom" but states that China's "laws and regulations clearly prohibit the spread of information that contains contents subverting state power, undermining national unity

[or] infringing upon national honor and interests." The Great Firewall of China, as the collection of state blocking tools is known, is nothing less than the guardian of Chinese statehood: "Within Chinese territory the Internet is under the jurisdiction of Chinese sovereignty. The Internet sovereignty of China should be respected and protected." This type of unabashed and unapologetic approach to censorship would naturally appeal to states with strong authoritarian streaks, as well as states with particularly impressionable or very homogenous populations (who would fear the incursion of outside information on an emotional level).

Next, there are the sheepish Internet filterers: Turkey has taken a much more subtle approach than China, and has even shown responsiveness to public demands for Internet freedom, but nevertheless its online censorship policies continue with considerable obfuscation. The Turkish government has had an uneasy relationship with an open Internet, being far more tolerant than some of its regional neighbors but much more restrictive than its European allies. It is impossible to get a completely unfiltered connection to the Internet in Turkey— an important distinction between Turkey and Western countries. YouTube was blocked by Turkish authorities for more than two years after the company refused to take down videos that officials claimed denigrated the country's founder, Mustafa Kemal Atatürk. (In keeping with a 1951 law that criminalizes public insults to Atatürk, YouTube agreed to block the videos for the Turkish audience, but the government wanted them removed globally from the platform worldwide.) This ban was highly visible, but subsequent censorship has been more covert: Some eight thousand websites have been blocked in Turkey without public notice or official government confirmation.

The sheepish model is popular with governments that struggle to strike a balance between divergent beliefs, attitudes and concerns within their population. But by pursuing this path, the government itself can become the enemy if it goes too far, or if its machinations are exposed. To give a recent example from Turkey: In 2011, the government announced a new nationwide Internet filtering policy featuring a four-tier system of censorship, in which citizens would have to choose the level of filtering they wanted (from the most to least restrictive: "child," "family," "domestic" and "standard" levels). The Information and Communications Technologies Authority (known by its initials in

Turkish as BTK) said the scheme was intended to protect minors and promised that people who chose the "standard" level would encounter no censorship. Many people skeptical of BTK's record on transparency balked. In fact, the plan generated such an outcry among the population that thousands of people in more than thirty cities around Turkey took to the streets to protest the proposed changes.

Under pressure, the government dialed back its plan, ultimately instituting just two content filters—"child" and "family"—which users could adopt voluntarily. But the controversy didn't end there. Media-freedom groups reported that their own tests of the censorship system revealed a more aggressive filtering framework than BTK would admit. In addition to the expected banned terms having to do with pornography or violent content, they found that ordinary news websites, content that was culturally liberal or Western (e.g., anything including the word "gay," or information about evolution) and keywords related to the Kurdish minority were all blocked under the new system. Some activists argued that blocking information about Kurdish separatist organizations with the "child" filter was evidence of the state's nefarious intent; the international media watchdog group Reporters Without Borders called the Turkish policy "backdoor censorship."

The Turkish government responded to some of the public concerns about the new system. When a Turkish newspaper reported that educational websites about scientific evolution were blocked while content from a prominent Turkish creationist were not, the authorities eliminated the block immediately. But there is little to no transparency around what content is censored under these policies, so the government is forced to react only when such discrepancies are brought to light by citizens. The sheepish model of Internet filtering, then, combines a government's ability to evade accountability with its willingness to take constructive action when pressure mounts. This approach would appeal to countries with growing civil societies but strong state institutions, or for governments without reliable bases of support but enough concentrated power to make such unilateral decisions.

The third approach, politically and culturally acceptable filtering, is employed by states as diverse as South Korea, Germany and Malaysia. This is limited and selective filtering around very specific content, based in law, with no attempt to hide the censorship or the motivations behind it. Outliers within the population might grumble, but the majority of citizens often agree with the filtering policies for reasons of security or public well-being. In South Korea, for example, the National Security Law expressly criminalizes public expressions of support for North Korea in both physical and virtual space. The South Korean government regularly filters Internet content affiliated with its northern neighbor—in 2010 it was reported that the government blocked some forty websites associated with or supportive of the North Korean regime, took down a dozen accounts with potential ties to Pyongyang on social-networking sites like Facebook and Twitter, and forced website administrators to delete more than forty thousand pro–North Korea blog posts.

Germany has strong anti-hate-speech laws that make Holocaust denial and neo-Nazi rhetoric illegal, and consequently the government blocks websites within Germany that express those views. And Malaysia, despite promising its citizens that it would never censor the Internet—going so far as to codify it in its Bill of Guarantees— abruptly blocked access to file-sharing sites like Megaupload and the Pirate Bay in 2011, claiming that the sites were in violation of another law, the country's Copyright Act of 1987. In a statement, the Malaysian Communications and Multimedia Commission defended the move, writing, "Compliance with the law is not to be construed as censorship." Many Malaysians disagreed, but the block remained politically and legally acceptable.

Of the three models, activists will pray that the third approach becomes the norm for states around the world, but this seems unlikely; only countries with highly engaged and informed populations will need to be this transparent and restrained. Since most governments will make such decisions before their citizens become fully connected, they will feel little incentive to proactively promote the kind of free and open Internet exhibited by countries in the "politically acceptable model."

The trends we see today will continue in ways that are, for the most part, fairly predictable. All governments will feel as if they're fighting a losing battle against an endlessly replicating and changing Internet, and balkanization will emerge as a popular mechanism to address this challenge. The next stage in the process for many states will be collective editing, states forming communities of interest to edit the web together, based on shared values or geopolitics. Collective action—be it in the physical or virtual world—will be a logical move for many states that find they lack the resources, the reach or the capability to influence vast territories. And even with balkanization, cyberspace is still a lot of ground to cover, so just as some states leverage each other's military resources to secure more physical ground, so too will states form alliances to control more virtual territory. For larger states, collaborations will legitimize their filtering efforts and deflect some unwanted attention (the "look, others are doing it too" excuse). For smaller states, alliances along these lines will be a low-cost way to curry favor with bigger players and simultaneously gain some useful technical skills and capacity that they might lack at home.

Collective editing may start with basic cultural agreements and shared antipathies among states, such as what religious minorities they dislike, how they view other parts of the world or what their cultural perspective is on historical figures like Vladimir Lenin, Mao Zedong or Mustafa Kemal Atatürk. In the online world, shared cultural and normative sensibilities create a gravitational pull among states, including those who might not otherwise have reason to band together. Larger states are less likely to engage in this than smaller ones—they already have the technical capabilities—so it will be a fleet of smaller states, pooling their resources, that will find this method useful. If some member countries in the Commonwealth of Independent States (CIS), an association of former Soviet states, became fed up with Moscow's insistence on standardizing the Russian language across the region, they could join together to censor all Russian-language content from their national Internets and thus limit their citizens' exposure to Russia altogether.

Ideology and religious morals are likely to be the strongest drivers of these collaborations. They are already the strongest drivers of censorship today. Imagine if a group of deeply conservative Sunni-majority countries—say, Saudi Arabia, Yemen, Algeria and Mauritania—formed an online alliance around their common values and strategic needs and decided to build a "Sunni Web." While technically this Sunni Web would still be part of the larger Internet, it would become the main source of information, news, history and activity for citizens living in these countries. For years, the development and spread of the Internet was highly determined by its English-only language standard, but the continued implementation of internationalized domain names (IDN), which allow people to use and access domain names written in non-Roman alphabet characters (e.g., http://‏مثال.إختبار‏), is changing this. The creation of a Sunni Web—indeed, all nationalized Internets—becomes more likely if its users can access a version of the Internet in their own language and script.

Within the Sunni Web, depending on who participated and who led its development, the Internet could be sharia-complicit: e-commerce and e-banking would look different, since no one would be allowed to charge interest; religious police might monitor online speech, working together with domestic law enforcement to report violations; websites with gay or lesbian content would be uniformly blocked; women's movements online might somehow be curtailed; and ethnic and religious minority groups might find themselves closely monitored, restricted or even excluded. In this scenario, how possible it would be for a local tech-savvy citizen to circumvent this Internet and reach the global World Wide Web depends on which country he lived in: Mauritania might not have the desire or capacity to stop him, but Saudi Arabia probably would. If the Mauritanian government became concerned that its users were bypassing the Sunni Web, on the other hand, surely one of its new digital partners could help it build higher fences. Within collective editing alliances, the less paranoid states would allow their populations to access both versions of the Internet (somewhat like an opt-in parental control for television), betting on user preference for safe and uniquely tailored content instead of using brute force.

There will be some instances where autocratic and democratic nations edit the web together. Such a collaboration will typically happen when a weaker democracy is in a neighborhood of stronger autocratic states that coerce it to make the same geopolitical compromises online that it makes in the physical world. This is one of the rare instances where physical proximity actually matters in virtual affairs. For example, Mongolia is a young democracy with an open Internet, sandwiched between Russia and China—two large countries with their own unique and restrictive Internet policies. The former Mongolian prime minister Sukhbaatar Batbold explained to us that he wants

Mongolia, like any country, to have its own identity. This means, he said, it must have good relations with its neighbors to keep them from meddling in Mongolian affairs. "We respect that each country has chosen for itself its own path in development," he said. With China, "we have an understanding where we stay out of Tibet, Taiwan and Dalai Lama issues, and they do not interfere with our issues. The same applies with Russia, with which we have a long-standing relationship."

A neutral stance of noninterference is more easily sustainable in the physical world. Virtual space significantly complicates this model because online, it's people who control the activity. People sympathetic to opposition groups and ethnic minorities within China and Russia would look at Mongolia as an excellent place to congregate. Supporters of the Uighurs, Tibetans or Chechen rebels might seek to use Mongolia's Internet space as a base from which to mobilize, to wage online campaigns and build virtual movements. If that happened, the Mongolian government would no doubt feel the pressure from China and Russia, not just diplomatically but because its national infrastructure is not built to withstand a cyber assault from either neighbor. Seeking to please its neighbors and preserve its own physical and virtual sovereignty, Mongolia might find it necessary to abide by a Chinese or Russian mandate and filter Internet content associated with hot-button issues. In such a compromise, the losers would be the Mongolians, whose online freedom would be taken away as a result of self-interested foreign powers with sharp elbows.

Not all states will look to collaborate with others during the balkanization process, but the end result just the same will be a jumble of national Internets and virtual borders. The trend toward globalized platforms like Facebook and Google creates a system for technology that is more likely to spread, which will mean a broader distribution of engineering tools that people can use to build their own online structures. Without state regulation that inhibits innovation, this growth trend will happen very rapidly. In the early stages, users won't realize when they are on another country's Internet because the experience will be seamless, as it is today. While states work to carve out their autonomy in the online world, most users will experience very little change.

That homeostasis, however, will not last. What started as the World

Wide Web will begin to look more like the world itself, full of internal divisions and divergent interests. Some form of visa requirement will emerge on the Internet. This could be done quickly and electronically, as a method to contain the flow of information in both directions, requiring that users register and agree to certain conditions to access a country's Internet. If China decides that all outsiders need to have a visa to access the Chinese Internet, citizen engagement, international business operations and investigative reporting will all be seriously affected. This, along with internal restrictions of the Internet, suggests a twenty-first-century equivalent of Japan's famous *sakoku* ("locked country") policy of near-total isolation enacted in the seventeenth century.

Some states may implement visa requirements as both a monitoring tool for international visitors and as a revenue-generating exercise—a very small fee would be charged automatically upon entering a country's virtual space, even more if one's online activities (which the government could track by cookies and other tools) violated the terms of the visa. Virtual visas would appear in response to security threats related to cyber attacks; if your IP came from a blacklisted country, you would encounter heightened screenings and monitoring.

Some states, however, would make a public show of not requiring visas to demonstrate their commitment to open data and to encourage other states to follow their example. In 2010, Chile became the first country in the world to approve a law that guarantees net neutrality. About half of Chile's 17 million people are online today, and as the country continues to develop its technological infrastructure, public statements like this will no doubt endear Chile to other governments that support its forward-looking communication policies. Countries coming online now will weigh the Chilean model against others. They might be asked to sign no-visa commitments with other states in order to build trade relations around e-commerce and other online platforms, like a Schengen Agreement (Europe's borderless zone) for the virtual world.

Under conditions like these, the world will see its first Internet asylum seeker. A dissident who can't live freely under an autocratic Internet and is refused access to other states' Internets will choose to seek physical asylum in another country to gain virtual freedom on its Internet. There could be a form of interim virtual asylum, where the host country would share sophisticated proxy and circumvention tools that would allow the dissident to connect outside. Being granted virtual asylum could be a significant first step toward physical asylum, a sign of trust without the full commitment. Virtual asylum would serve as an extra layer of vetting before the physical asylum case reached the courts.

Virtual asylum will not work, however, if the ultimate escalation occurs: the creation of an alternative domain name system (DNS), or even aggressive and ubiquitous tampering with it to advance state interests. Today, the Internet as we know it uses the DNS to match computers and devices to relevant data sources, translating IP addresses (numbers) into readable names and vice versa. The robustness of the Internet depends on all computers and networks' using the same official DNS root (run by the Internet Corporation for Assigned Names and Numbers, or ICANN), which contains all the top-level domains that appear as suffixes on web addresses—.edu, .com, .net and others.

But there are alternative DNS roots in existence, operating in parallel with the Internet but not attached to it. Within tech circles, most believe that the creation of an alternative DNS would go against everything the Internet represents and was built to do: namely, share information freely. No government has yet achieved an alternative system,* but if a government succeeded in doing so, it would effectively unplug its population from the global Internet and instead offer only a closed, national intranet. In technical terms, this would entail creating a censored gateway between a given country and the rest of the world, so that a human proxy could facilitate external data transmissions when absolutely necessary—for matters involving state resources, for instance.

For the population, popular proxy measures like VPNs and Tor would no longer have any effect because there wouldn't be anything to connect to. It's the most extreme version of what technologists call a walled garden. On the Internet, a walled garden refers to a browsing environment that controls a user's access to information and services online. (This concept is not limited to discussions of censorship; it has

---

* Smaller incidents, however, do suggest that governments are capable and perhaps comfortable manipulating DNS routing on occasion. More than a few times, Google's web address has mysteriously directed people to www.Baidu.com, China's local search competitor.

deep roots in the history of Internet technology: AOL and Compu-Serve, Internet giants for a time, both started as walled gardens.) For the full effect of disconnection, the government would also instruct the routers to fail to advertise the IP addresses of websites—unlike DNS names, IP addresses are immutably tied to the sites themselves—which would have the effect of putting those websites on a very distant island, utterly unreachable. Whatever content existed on this national network would circulate only internally, trapped like a cluster of bubbles in a computer screen saver, and any attempts to reach users on this network from the outside would meet a hard stop. With the flip of a switch, an entire country would simply disappear from the Internet.

This is not as crazy as it sounds. It was first reported in 2011 that the Iranian government's plan to build a "halal Internet" was under way, and more than a year later it seemed that the official launch was imminent. The regime's December 2012 launch of *Mehr,* its own version of YouTube with "government-approved videos," added yet another data point that the regime was serious about the project. Details of the plan remained hazy, but according to Iranian government officials, in the first phase the national "clean" Internet would exist in tandem with the global Internet for Iranians (heavily censored as it is), then it would come to replace the global Internet altogether. This would entail moving all the "halal" websites to a particular block of IP addresses, which would make it trivially easy to filter out websites that are outside the halal block. The government and affiliated institutions would provide the content for the national intranet, either gathering it from the global web and scrubbing it, or creating it manually. All activity on the network would be closely monitored, facilitated by the government's top-level infrastructure control and agency over software (something Iranian officials are very concerned about, judging from a 2012 ban on the import of foreign computer security software). Iran's head of economic affairs told the country's state-run news agency that they hoped their halal Internet would come to replace the web in other Muslim countries, too—at least those with Farsi speakers. Pakistan has pledged to build something similar.

It is possible that Iran's threat is merely a hoax. How exactly the state intends to proceed with this project is unclear both technically and politically. How would it avoid enraging the sizable chunk of its population that has access to the Internet? Some believe it would be impossible to fully disconnect Iran from the global Internet because of its broad economic reliance on external connections. Others speculate that, if it wasn't able to build an alternative root system, Iran could pioneer a dual-Internet model that other repressive states would want to follow. Whichever route Iran chooses, if it is successful in this endeavor, its halal Internet would surpass the Great Firewall of China as the single most extreme version of information censorship in history. It would change the Internet as we know it.

Virtual Multilateralism

In parallel with these balkanization efforts, we will see the rise of virtual multilateralism based on ideological or political solidarity, involving both states and corporations working together in official alliances. States like Belarus, Eritrea, Zimbabwe and North Korea—authoritarian, with strong personality cults and a pariah status elsewhere in the world—would have little to lose by joining an autocratic cyber union, where censorship and monitoring strategies and technologies could be shared. As these countries collaborated to build virtual-age police states, it would become increasingly difficult for Western companies, from a public-relations standpoint, to conduct business there, even if it was legal. This would create space for non-Western companies, whose shareholders may have fewer qualms and who are used to working in similar environments, to play a more active business role within a network of autocratic states.

It's no accident, for example, that the company that owns 75 percent of North Korea's only official mobile network, Koryolink, is the Egyptian telecom Orascom, a firm that thrived under the long reign of Hosni Mubarak. (The other 25 percent is owned by North Korea's Ministry of Posts and Telecommunications.) For North Korean subscribers, Koryolink service is a walled garden, a highly limited platform that allows for only basic functionality. Koryolink users can't make or receive international calls; nor can they access the Internet. (Some people can access the North Korean intranet, an odd pastiche of online content, mostly propaganda, that government officials transfer over from the Internet.) Local phone calls and text messages are almost certainly monitored, and *The Economist* reported that the network is already a platform for the dissemination of government propaganda, with the North Korean daily *Rodong Sinmun* sending users the latest news by text message. While it is not officially a requirement, most people are "encouraged" to pay their phone bills in euros (which are unofficially in circulation), a tall order for most North Koreans. Even so, the demand for phones was so great that adoption soared in the country, leaping from three hundred thousand subscribers to more than a million within an eighteen-month period ending in early 2012. Koryolink's gross operating margin of 80 percent means big business for Orascom.

In Iran, following a very public crackdown on the country's green movement in 2009, Western technology companies like Ericsson and Nokia Siemens Networks (NSN) sought to distance themselves from the regime. In their absence, the Chinese telecommunications giant Huawei swept in and seized the opportunity to dominate the large (and state-controlled) Iranian mobile market. While its Western predecessors faced a backlash at home for selling products to the Iranian government that were used to track and suppress democracy activists, Huawei actively promoted its products in an authoritarian-friendly light. Its catalog was unapologetic, according to a story in *The Wall Street Journal,* with products like location-based tracking equipment for law enforcement (recently purchased by Iran's largest mobile operator) and a censorship-friendly mobile news service. Huawei's favorite domestic partner in Iran, Zaeim Electronic Industries Co., is also the favorite of government branches, including the Revolutionary Guards and the office of the president.

Officially, Huawei claims to offer Zaeim only "commercial public-use products and services," but according to *The Wall Street Journal,* in off-the-record pitch meetings with Iranian officials, Huawei made clear its expertise in information censorship, mastered in China. (Huawei published a press release shortly after the story's publication denying several of its assertions, and a month later stated that it would "voluntarily restrict" its business operations in Iran due to the "increasingly complex situation.")

In response to these collaborations between autocratic countries, democratic states will want to build similar alliances and public-private partnerships to promote a more open Internet with greater political, economic and social freedom. One goal will be to contain the spread of highly restrictive filtering and monitoring technologies to countries with low but growing Internet penetration. This could manifest itself in many different strategies, including bilateral assistance packages with specific preconditions and making an open Internet a premier policy objective for a country's ambassadors. There could also be transnational campaigns to change the international legal framework around free expression and open-source software. The shared, "bigger picture" goals of these states—access to information, freedom of expression, and transparency—would trump the minor policy or cultural differences between them, creating a kind of revived Hanseatic League of connectivity. The Hanseatic League wielded collective power across Northern Europe from the thirteenth century through the fifteenth through its economic alliances between adjacent city-states; its contemporary equivalent could be based on similar principles of mutual assistance but in a far larger, globalized version. No longer will alliances rely so heavily on geography; everything is equidistant in virtual space. If Uruguay and Benin find cause to work together, it will be easier to do so than ever before.

Part of defending freedom of information and expression in the future will entail a new element of military aid. Training will include technical assistance and infrastructural support in lieu of tanks and tear gas—though the latter will probably remain part of the arrangement. What Lockheed Martin was to the twentieth century, technology and cyber-security companies will be to the twenty-first. Indeed, traditional defense-industry leaders like Northrop Grumman and Raytheon are already working with the U.S. government to develop cyber-capacity. Weapons manufacturers, airplane builders and other parts of the military-industrial complex might not lessen—conventional militaries will always require guns, tanks and helicopters—but big military operations, already heavily privatized, will carve out space in their budgets for technical assistance.

Development assistance and foreign aid will take on a digital dimension too, buoyed by these new multilateral alliances. The trade of foreign assistance for future influence won't change, but the components will. In a given developing country, one foreign power might be building roads, another investing in agriculture and a third building fiber networks and cell towers. In the digital age, modern technology becomes yet another tool for forging alliances with developing states; we shouldn't underestimate how important technological competency will be for these countries and their governments. The push for foreign aid in the shape of fast networks, modern devices, and cheap and plentiful bandwidth may come from the population, pressuring the governments to agree to the necessary preconditions. Whatever the impetus, future states in the developing world will make a long-term bet on connectivity and align their diplomatic relationships accordingly.

New alliances will form around commercial interests as well, particularly copyright and intellectual-property issues. As commerce moves increasingly to the online world, the dynamics around copyright enforcement will lead to another layer of virtual alliances and adversaries. Most copyright and intellectual-property laws are still centered on the notion of physical goods, and there are divergent attitudes about whether theft or piracy of online goods (movies, music and other content) are equivalent to the theft of physical versions of those same items. In the future, states will begin to wade more deeply into legal battles over copyright and intellectual property because the health of their commercial sectors will be at stake.

There have been multiple international agreements dealing with copyright laws: the Berne Convention of 1886, which requires mutual recognition of the copyrights of other signatory states; the Agreement on Trade-Related Aspects of Intellectual Property Rights of 1994, which set the minimum standards for intellectual-property rights in World Trade Organization (WTO) states; and the World Intellectual Property Organization (WIPO) Copyright Treaty of 1996, which protects information-technology copyrights against infringement. The laws that govern copyright around the world are generally the same. But each country is responsible for enforcement within its borders, and not all countries are equally vigilant. Given the ease with which information crosses borders, people who pirate copyrighted material are typically able to find virtual safe havens in countries with less stringent regulation.

The great concern among intellectual-property watchers in the technology world is China. Because it is a signatory to the conventions above, technically it is bound to the same standards as other countries, including the United States. At the Asia-Pacific Economic Cooperation (APEC) CEO Summit in 2011, then Chinese president Hu Jintao privately told a small group of business leaders that China would "fully implement all of the intellectual property laws as required by the World Trade Organization and modern Western practices." We attended this meeting, and as we filed out of the room after President Hu's comments, the American business contingent clearly expressed skepticism toward his claim. And with good reason: It's estimated that U.S. companies lost approximately $3.5 billion in 2009 alone because of pirated music recordings and software from China, and that 79 percent of all copyright-infringing goods seized in the United States were produced in China. Clearly, it's not the absence of laws that contribute to this problem, but their lack of enforcement. Officially, it's against Chinese law to produce counterfeit goods or to copy intellectual property for profit, but in practice, officials are discouraged from pursuing criminal prosecution of these crimes; violators are allowed to keep their profits. Moreover, the fines for violating the laws are too low and too irregularly issued to be effective in deterring such behavior, and corruption at local and regional levels encourages officials to turn the other way and ignore repeated violations.

China is by no means the only state unwilling or unable to enforce international intellectual-property norms. Russia, India and Pakistan have all been singled out for their equally dismal enforcement of these laws. Israel and Canada aren't normally considered hotbeds of copyright infringement, but neither country has fully implemented the standards and laws of the WIPO, making them a haven for Internet piracy. And within the group of states that do have strong protections for intellectual-property rights, there are usually significant and exploitable differences in interpretation. For example, the notion of fair use (as the United States terms it) or fair dealing (as the British do), which allows for the limited use of copyrighted material without consent from the copyright holder, is far more tightly controlled in the European Union than it is in either the United States or the United Kingdom.

Virtual Statehood

One of our recurring themes is that in the virtual world, size matters less. Technology empowers all parties, and allows smaller actors to have outsized impacts. And those actors need not be known or official. To wit, we believe it's possible that virtual states will be created and will shake up the online landscape of physical states in the future.

There are hundreds of active violent and nonviolent secessionist movements in the world today, and this is unlikely to change in the future. A large portion of the movements are motivated by perceived ethnic or religious discrimination, and shortly we will discuss how physical discrimination and persecution of these groups will play out online, changing shape but not intent. In the physical world, it's not uncommon for persecuted groups to be subject to different laws and vulnerable to indeterminate detention, extrajudicial killings, the absence of due process, and all manner of restrictions on their civil and human liberties, and most of these tactics will find their way online, aided significantly by technology that helps regimes monitor, harass and target their restive minority populations.

Hounded in both the physical and virtual worlds, groups that lack formal statehood may choose to emulate it online. While not as legitimate or useful as actual statehood, the opportunity to establish sovereignty virtually could prove to be, in the best cases, a meaningful step toward official statehood, or in the worst cases, an escalation that further entrenches both sides in a messy civil conflict. The Kurdish populations in Iran, Turkey, Syria and Iraq—the four countries where they are most concentrated—might build a Kurdish web as a way to carve out a sort of virtual independence. Iraqi Kurdistan is already quasi-autonomous, so the efforts could begin there. Kurds could establish a top-level domain (e.g., www.yahoo.com.krd), with "krd" standing for Kurdistan, by registering a new domain and basing the servers in a neutral or supportive country. Then they'd build upon that.

Virtual statehood would be much more than just a gesture and a domain name. Additional projects could also develop a distinct Kurdish presence online. With enough effort, the Kurdish web could become a robust version of other countries' Internet, in the Kurdish language, of course. From there, Kurdish or sympathetic engineers could build applications, databases and other online destinations that not only support the Kurdish cause but actually facilitate it. The virtual Kurdish community could hold elections and set up ministries to deliver basic public goods. They could even use a unique online currency. The virtual minister of information would manage the data flow to and from the online Kurdish "citizens." The minister of the interior would focus on preserving the security of the virtual state and protecting it from cyber attack. The foreign minister would engage in diplomatic relations with other, actual states. The economic and trade minister would promote e-commerce between Kurdish communities and outside economic interests.

Just as secessionist efforts to move toward physical statehood are typically resisted strongly by the host state, such groups would face similar opposition to their online maneuvers. The creation of a virtual Chechnya might cement ethnic and political solidarity among its supporters in the Caucasus region, but it would no doubt worsen relations with the Russian government, which would consider such a move a violation of its sovereignty. The Kremlin might well respond to virtual provocation with a physical crackdown, rolling in tanks and troops to quell the stirrings in Chechnya.

For the Kurds, who stretch across several countries, this risk would be even more pronounced, as a Kurdish virtual-statehood campaign would be met with resistance from the entire neighborhood, some of whom lack Kurdish populations but would fear a destabilizing effect. No effort would be spared to destroy the Kurdish virtual institutions through low-grade cyber-meddling and espionage, like cyber attacks, disinformation campaigns and infiltration. The populations on the ground would surely bear the brunt of the punishment. The governments would be aided, of course, by the massive amounts of data that these citizens produced, so finding the people involved or supportive of virtual statehood would be easy. Very few secessionist movements have the level of resources and international support that would be required to match this level of counterattack.

Declaring virtual statehood would become an act of treason, not just in restive regions but almost everywhere. It's simply too risky an avenue to leave open. The concept of virtual institutions alone could breathe new life into secessionist groups that have tried and failed to produce concrete outcomes through violent means, like the Basque separatists in Spain, the Abkhaz nationalists in Georgia or the Moro

Islamic Liberation Front in the Philippines. One failed or unsuitable effort could also break the experiment altogether. If, for example, the lingering supporters of the Texas secession movement rallied together to launch a virtual Republic of Texas, and they were met with derision, the concept of virtual statehood might be sullied for some time. How successful these virtual statehood claims would be (what would constitute success, in the end?) remains to be seen, but the fact that this will be feasible says something significant about the diffusion of state power in the digital age.

Digital Provocation and Cyber War

No discussion on the future of connected states would be complete without a look at the worst things they'll do to each other: namely, launch cyber wars. Cyber warfare is not a new concept, nor are its parameters well established. Computer security experts continue to debate how great the threat is, what it looks like and what actually constitutes an act of cyber war. For our purposes, we'll use the definition of cyber warfare offered by the former U.S. counterterrorism chief Richard Clarke: actions by a nation-state to penetrate another nation's computers or networks for the purposes of causing damage or disruption.*

Cyber attacks—including digital espionage, sabotage, infiltration and other mischief—are, as we established earlier, very difficult to trace and have the potential to inflict serious damage. Both terrorist groups and states will make use of cyber-war tactics, though governments will focus more on information-gathering than outright destruction. For states, cyber war will primarily meet intelligence objectives, even if the methods employed are similar to those used by independent actors looking to cause trouble. Stealing trade secrets, accessing classified information, infiltrating government systems, disseminating misinformation—all traditional activities of intelligence agencies—will make up the bulk of cyber attacks between states in the future.

---

* We distinguish between "cyber attack" and "cyber terrorism" by looking at the individual or entity behind the attack and assessing motives. The two, however, may manifest themselves in very similar ways, such as economic espionage.

Others fundamentally disagree with us on this point, predicting instead that states will seek to destroy their enemies by heavy-handed methods like cutting off power grids remotely or crashing stock markets. In October 2012, the U.S. secretary of defense, Leon Panetta, warned, "An aggressor nation . . . could use these kinds of cyber tools to gain control of critical switches. They could derail passenger trains, or even more dangerous, derail passenger trains loaded with chemicals. They could contaminate the water supply in major cities, or shut down the power grid across large parts of the country." We tend to take the optimist's perspective (at least when it comes to states) and say that such escalations, while possible, are highly unlikely, if only because the government that first starts this trend would itself become a target as well as set a precedent that even the most erratic regimes would be cautious to approach.

It's fair to say that we're already living in an age of state-led cyber war, even if most of us aren't aware of it. Right now, the government of a foreign country could be hacking into your government's databases, crashing its servers or monitoring its conversations. To outside observers, our current stage of cyber war might seem benign (indeed, some might contend that it's not really "war" anyway, as per the classical Clausewitzian framework of "war as a continuation of policy by other means"). Government-backed engineers might be trying to infiltrate or shut down the information systems of companies and institutions in other countries, but no one is getting killed or wounded. We've seen so little spillage of these cyber wars into the physical world that for civilians, a cyber attack seems more an inconvenience than a threat, like an attack of the common cold.

But those who underestimate the threat of cyber war do so at their peril. While not all the hype surrounding cyber war is justified, the risks are real. Cyber attacks are occurring with greater frequency and more precision with each passing year. The increasing entwining of our lives with digital-information systems leaves us more vulnerable with each click. And as many more countries come online in the near future, those vulnerabilities will only expand and become more complicated.

A cyber attack might be the state's perfect weapon: powerful, customizable and anonymous. Tactics like hacking, deploying computer worms or Trojan horses and other forms of virtual espionage present states with more reach and more cover than they would have with traditional weapons or intelligence operations. The evidence trails they leave are cold, providing perpetrators with effective camouflage and severely limiting the response capability of the victims. Even if an attack could be traced back to a particular region or town, identifying the responsible parties is nearly impossible. How can a country determine an appropriate response if it can't prove culpability? According to Craig Mundie, Microsoft's chief research and strategy officer and a leading thinker in Internet security, the lack of attribution—one of our familiar themes—makes this a war conducted in the dark, because "it's just much harder to know who took the shot at you." Mundie calls cyber-espionage tactics "weapons of mass disruption." "Their proliferation will be much faster, making this a much stealthier kind of conflict than has classically been determined as warfare," he said.

States will do things to each other online that would be too provocative to do off-line, allowing conflicts to play out in the virtual battleground while all else remains calm. The promise of near-airtight anonymity will make cyber attacks an attractive option for countries that don't want to appear overtly aggressive but remain committed to undermining their enemies. Until the world's technical experts get better at determining the origin of cyber attacks and the law is able to hold perpetrators to account, many more states will join in on the activities we see today. Blocks of states that are already gaining connectivity and technical capacity, in Latin America, Southeast Asia and the Middle East, will begin launching their own cyber attacks soon, if only to test the waters. Even those who lack indigenous technical skills (e.g., local engineers and hackers) will find ways to get the tools they need.

Let's consider a few recent examples to better illustrate the universe of cyber warfare. Perhaps the most famous is the Stuxnet worm, which was discovered in 2010 and was considered the most sophisticated piece of malware ever revealed, until a virus known as Flame, discovered in 2012, claimed that title. Designed to affect a particular type of industrial control system that ran on the Windows operating system, Stuxnet was discovered to have infiltrated the monitoring systems of Iran's Natanz nuclear-enrichment facility, causing the centrifuges to abruptly speed up or slow down to the point of self-destruction while simultaneously disabling the alarm systems. Because the Iranian systems were not linked to the Internet, the worm must have been uploaded directly, perhaps unwittingly introduced by a Natanz employee on a USB flash drive. The vulnerabilities in the Windows systems were subsequently patched up, but not after causing some damage to the Iranian nuclear effort, as the Iranian president, Mahmoud Ahmadinejad, admitted.

Initial efforts to locate the creators of the worm were inconclusive, though most believed that its target and level of sophistication pointed to a state-backed effort. Among other reasons, security analysts unpacking the worm (their efforts made possible because Stuxnet had escaped "into the wild"—that is, beyond the Natanz plant) noticed specific references to dates and biblical stories in the code that would be highly symbolic to Israelis. (Others argued that the indicators were far too obvious, and thus false flags.) The resources involved also suggested government production: Experts thought the worm was written by as many as thirty people over several months. And it used an unprecedented number of "zero-day" exploits, malicious computer attacks exposing vulnerabilities (security holes) in computer programs that were unknown to the program's creator (in this case, the Windows operating system) before the day of the attack, thus leaving zero days to prepare for it. The discovery of one zero-day exploit is considered a rare event—and exploited information can be sold for hundreds of thousands of dollars on the black market—so security analysts were stunned to discover that an early variant of Stuxnet took advantage of *five*.

Sure enough, it was revealed in June 2012 that not one but two governments were behind the deployment of the Stuxnet worm. Unnamed Obama administration officials confirmed to the *New York Times* journalist David E. Sanger that Stuxnet was a joint U.S. and Israeli project designed to stall and disrupt the suspected Iranian nuclear-weapons program.* Initially green-lit under President George W. Bush, the initiative, code-named Olympic Games, was carried into the next administration and in fact accelerated by President Obama, who personally authorized successive deployments of this cyber weapon. After building the malware and testing it on functioning replicas of the Natanz plant built in the United States—and discovering that it

---

* When we asked the former Israeli intelligence chief Meir Dagan about the collaboration, his only comment was, "Do you really expect me to *tell* you?"

could, in fact, cause the centrifuges to break apart—the U.S. government approved the worm for deployment. The significance of this step was not lost on American officials.* As Michael V. Hayden, the former CIA director, told Sanger, "Previous cyberattacks had effects limited to other computers. This is the first attack of a major nature in which a cyberattack was used to effect physical destruction. Somebody crossed the Rubicon."

When the Flame virus was discovered two years later, initial reports from security experts suggested that it was unconnected to Stuxnet; it was much larger, used a different programming language and operated differently, focusing on covert data-gathering instead of targeting centrifuges. It was also older—analysts found that Flame had been in existence for at least four years by the time they discovered it, which means it predated the Stuxnet worm. And Sanger reported that American officials denied that Flame was part of the Olympic Games project. Yet less than a month after the public revelations about these cyber weapons, security experts at Kaspersky Lab, a large Russian computer-security company with international credibility, concluded that the two teams that developed Stuxnet and Flame did, at an early stage, collaborate. They identified a particular module, known as Resource 207, in an early version of the Stuxnet worm that clearly shares code with Flame. "It looks like the Flame platform was a kick-starter of sorts to get the Stuxnet project going," a senior Kaspersky researcher explained. "The operations went separate ways, maybe because Stuxnet code was mature enough to be deployed in the wild. Now we are 100 percent sure that the Stuxnet and Flame groups worked together."

Though Stuxnet, Flame and other cyber weapons linked to the United States and Israel are the most advanced known examples of state-led cyber attacks, other methods of cyber warfare have already been used by governments around the world. These attacks needn't be limited to highly consequential geopolitical issues; they can be

---

* Larry Constantine, a professor at the University of Madeira, in Portugal, challenges Sanger's analysis in a September 4, 2012, interview podcast with Steven Cherry, an associate senior editor at *IEEE Spectrum,* the magazine of the Institute of Electrical and Electronics Engineers, arguing that it is technically impossible for Stuxnet to have spread in the manner that Sanger described (e.g., Stuxnet could spread only over a LAN—local area network—not the Internet). Our view is that Constantine's argument has enough validity to at least warrant debate.

deployed to harass a disliked fellow state with equal panache. Following a diplomatic fight in 2007 over the Estonian government's decision to remove a Russian World War II memorial in its capital, Tallinn, a mass of prominent Estonian websites, including those of banks, newspapers and government institutions, were abruptly struck down by a distributed denial of service (DDoS) attack. Estonia is often called the most wired country on Earth, because almost every daily function of the state (and nearly all of its citizens) employs online services, including e-government, e-voting, e-banking and m-parking, which allows drivers to pay for their parking with a mobile device. Yet the country that gave the world Skype suddenly found itself paralyzed due to the efforts of a group of hackers. The systems came back online, and the Estonians immediately suspected their neighbor Russia—the Estonian foreign minister, Urmas Paet, accused the Kremlin directly—but proving culpability was not possible. NATO and European Commission experts were unable to find evidence of official Russian government involvement. (The Russians, for their part, denied the charges.)

Some questions that arise—Was it an act of cyber warfare? Would it be if the Kremlin hadn't ordered it, but gave its blessing to the hackers who executed it?—remain unanswered. In the absence of attribution, victims of cyber attacks are left with little to go on, and perpetrators can remain safe from prosecution even if suspicion is heightened. (One year after the Estonian attacks, websites for the Georgian military and government were brought down by DDoS attacks, while the country was in a dispute with, you guessed it, Russia. The following year, Russian hackers targeted the Internet providers in Kyrgyzstan, shutting down 80 percent of the country's bandwidth for days. Some believe the attacks were intended to curb the Kyrgyz opposition party, which has a relatively large Internet presence, while others contend that the impetus was a failed investment deal, in which Russia had tried to get Kyrgyzstan to shut down the U.S. military base it hosted.)

Then there is the example of Chinese cyber attacks on Google and other American companies over the past few years. Digital corporate espionage is a rowdy subcategory of cyber warfare, a relatively new phenomenon that in the future will have a severe impact on relations between states as well as national economies. Google finds its systems under attack from unknown digital assailants frequently, which is why it spends so much time and energy building the most secure network and protections possible for Google users. In late 2009, Google detected unusual traffic within its network and began to monitor the activity. (As in most cyber attacks, it was more valuable to our cyber-security experts to temporarily leave the compromised channels open so that we could watch them, rather than shut them down imme-diately.) What was discovered was a highly sophisticated industrial attack on Google's intellectual property coming from China.

Over the course of Google's investigation, it gathered sufficient evi-dence to know that the Chinese government or its agents were behind the attack. Beyond the technical clues, part of the attacks involved attempts to access and monitor the Gmail accounts of Chinese human-rights activists, as well as the accounts of advocates of human rights in China based in the United States and Europe. (These attacks were largely unsuccessful.) In the end, this attack—which targeted not only Google but dozens of other publicly listed companies—was among the driving factors in Google's decision to alter its business position in China, resulting in the shutdown of its Google China operations, the end of self-censorship of Chinese Internet content, and the redirection of all incoming searches to Google in Hong Kong.

Today, only a small number of states have the capacity to launch large-scale cyber attacks—the lack of fast networks and technical tal-ent holds others back—but in the future there will be dozens more participating, either offensively or defensively. Many people believe that a new arms race has already begun, with the United States, China, Russia, Israel and Iran, among others, investing heavily in stockpil-ing technological capabilities and maintaining a competitive edge. In 2009, around the same time that the Pentagon gave the directive to establish United States Cyber Command (USCYBERCOM), then secretary of defense Robert Gates declared cyberspace to be the "fifth domain" of military operations, alongside land, sea, air and space. Perhaps in the future the military might create the equivalent of the Army's Delta Force for cyberspace, or we could see the establishment of a department of cyber war with a new cabinet secretary. If this sounds far-fetched, think back to the creation of the Department of Homeland Security as a response to 9/11. All it takes is one big national episode to spur tremendous action and resource allocation on the part of the government. Remember, it was the United Kingdom's experi-ence with Irish terrorism that led to the establishment of closed-circuit television (CCTV) cameras in every corner of London, a move that was welcomed by much of the populace. Of course, some raised concerns about their every move on the streets being filmed and stored, but in moments of national emergency, the hawks always prevail over the doves. Postcrisis security measures are extremely expensive, with states having to act quickly and go the extra mile to assuage the concerns of their population. Some cyber-security experts peg the cost of the new "cyber-industrial complex" somewhere between $80 billion and $150 billion annually.

Countries with strong engineering sectors like the United States have the human capital to build their virtual weapons "in-house," but what of the states whose populations' technical potential is underdeveloped? Earlier, we described a minerals-for-technology trade for governments looking to build surveillance states, and it stands to reason that this type of exchange will work equally well if those states' attention turns toward its external enemies. Countries in Africa, Latin America and Central Asia will locate supplier nations whose technological investment can augment their own lackluster infrastructure. China and the United States will be the largest suppliers but by no means the only ones; government agencies and private companies from all over the world will compete to offer products and services to acquisitive nations. Most of these deals will occur without the knowledge of either country's population, which will lead to some uncomfortable questions if the partnership is later exposed. A raid on the Egyptian state security building after the country's 2011 revolution produced explosive copies of contracts with private outlets, including an obscure British firm that sold online spyware to the Mubarak regime.

For countries looking to develop their cyber-war capabilities, choosing a supplier nation will be an important decision, akin to agreeing to be in their "sphere of online influence." Supplier nations will lobby hard to gain a foothold in emerging states, since investment buys influence. China has been remarkably successful in extending its footprint into Africa, trading technical assistance and large infrastructure projects for access to resources and consumer markets, in no small part due to China's noninterference policy and low bids. Who, then, will those countries likely turn to when they decide to start building their cyber arsenal?

Indeed, we already see signs of such investments under the umbrella of science and technology development projects. Tanzania, a former socialist country, is one of the largest recipients of Chinese foreign direct assistance. In 2007, a Chinese telecom was contacted to lay some ten thousand kilometers of fiber-optic cable. Several years later, a Chinese mining company called Sichuan Hongda announced that it had entered into a $3 billion deal with Tanzania to extract coal and iron ore in the south of the country. Shortly thereafter, the Tanzanian government announced it had entered into a loan agreement with China to build a natural-gas pipeline for $1 billion. All across the continent, similar symbiotic relationships exist between African governments and big Chinese firms, most of which are state-owned. (State-owned enterprises make up 80 percent of the value of China's stock market.) A $150 million loan for Ghana's e-governance venture, implemented by the Chinese firm Huawei, a research hospital in Kenya, and an "African Technological City" in Khartoum all flow from the Forum on China-Africa Cooperation (FOCAC), a body established in 2000 to facilitate Sino-African partnerships.

In the future, superpower supplier nations will look to create their spheres of online influence around specific protocols and products, so that their technologies form the backbone of a particular society and their client states come to rely on certain critical infrastructure that the superpower alone builds, services and controls. There are currently four main manufacturers of telecommunications equipment: Sweden's Ericsson, China's Huawei, France's Alcatel-Lucent and Cisco in the United States. China would certainly benefit from large portions of the world using its hardware and software, because the Chinese government has dominating influence over what its companies do. Where Huawei gains market share, the influence and reach of China grow as well. Ericsson and Cisco are less controlled by their respective governments, but there will come a time when their commercial and national interests align and contrast with China's—say, over the abuse of their products by an authoritarian state—and they will coordinate their efforts with their governments on both diplomatic and technical levels.

These spheres of online influence will be both technical and political in nature, and while in practice such high-level relationships may not affect citizens in daily life, if something serious were to happen (like an uprising organized through mobile phones), which technology a country uses and whose sphere it's in might start to matter. Technology companies export their values along with their products, so it is absolutely vital who lays the foundation of connectivity infrastructure. There are different attitudes about open and closed systems, disputes over the role of government, and different standards of accountability. If, for example, a Chinese client state uses its purchased technology to persecute internal minority groups, the United States would have very limited leverage: Legal recourse would be useless. This is a commercial battle with profound security implications.

The New Code War

The logical conclusion of many more states coming online, building or buying cyber-attack capability and operating within competitive spheres of online influence is perpetual, permanent low-grade cyber war. Large nations will attack other large nations, directly and by proxy; developing nations will exploit their new capabilities to address long-standing grievances; and smaller states will look to have a disproportionately large influence, safe in the knowledge that they won't be held accountable because of the untraceable nature of their attacks. Because most attacks will be silent and slow-moving information-gathering exercises, they won't provoke violent retaliation. That will keep tensions on a slow burn for years to come. Superpowers will build up virtual armies within their spheres of influence, adding an important proxy layer to insulate them, and together they'll be able to produce worms, viruses, sophisticated hacks and other forms of online espionage for commercial and political gain.

Some refer to this as the upcoming Code War, where major powers are locked in a simmering conflict in one dimension while economic and political progress continues unaffected in another. But unlike its real-world predecessor, this won't be a primarily binary struggle; rather, the participation of powerful tech-savvy states including Iran, Israel and Russia will make it a multipolar engagement. Clear ideological fault lines will emerge around free expression, open data and liberalism. As we said, there will be little overt escalation or spillage into the physical world because none of the players would want to jeopardize their ongoing relationships.

Some classic Cold War attributes will carry over into the Code War, particularly those pertaining to espionage, because governments will largely view their new cyber-warfare capabilities as extensions of their intelligence agencies. Embedded moles, dead letter drops and other tradecraft will be replaced by worms, key-logging software, location-based tracking and other digital spyware tools. Extracting information from hard drives instead of from humans may reduce risk to traditional assets and their handlers, but it will introduce new challenges, too: Misinformation will remain a problem, and very sophisticated computers may give up secrets even less easily than people.

Another Cold War attribute—war by proxy—will see a revival in these new digital-age entanglements. On one hand, it could manifest in progressive alliances between states to counter dangerous non-state elements, where the cyber attack's lack of attribution provides political cover. The United States could covertly fund or train Latin American governments to launch electronic attacks on drug-cartel networks. On the other hand, war by digital proxy could lead to further misdirection and false accusations, with countries exploiting the lack of attribution for their own political or economic gain.

As with the Cold War, there will be little civilian involvement, awareness or direct harm, which deleteriously affects how states perceive the risks of such activities. States with ambition but a lack of experience in cyber warfare might go too far and unintentionally start a conflict that actually does harm their populations. Eventually, mutually-assured-destruction doctrines might emerge between states that stabilize these dynamics, but the multipolarity of the landscape promises to keep some measure of volatility in the system.

More important, there will be a great deal of room for error in the new Code War. The misperceptions, misdirection and mistakes that characterized the Cold War era will reappear with vigor as all participants go through the process of learning how to use the powerful new tools at their disposal. Given the additional layer of obfuscation that cyber attacks provide, it might end up being worse than the Cold War—even exploded missiles leave trails. Mistakes will be made by governments in deciding what to target and how, by victims who out of panic or anger retaliate against the wrong party, and by the engineers who construct these massively complicated computer programs.

With weapons this technically complex, it's possible that a rogue individual would install his own back door in the program—a means of access that bypasses security mechanisms and can be used remotely—which would remain unnoticed until he decided to use it. Or perhaps a user would unknowingly share a well-constructed virus in a way its creators did not intend, and instead of skimming information about a country's stock exchange, it would actually crash it. Or a dangerous program could be discovered that would bear several false flags (the digital version of bait) in the code, and this time the targeted country would decide to take action against the apparent source.

We've already seen examples of how the attribution problem of cyber attacks can lead to misdirection on a state level. In 2009, three waves of DDoS attacks crippled major government websites in both the United States and South Korea. When security experts reviewed the cyber attack, they found Korean language and other indicators that strongly suggested that the network of attacking computers, or botnet, began in North Korea. Officials in Seoul directly pointed their fingers at Pyongyang, the American media ran with the story and a prominent Republican lawmaker demanded that President Obama conduct a "show of force or strength" against North Korea in retaliation.

In fact, no one could *prove* where the attacks came from. A year later, analysts concluded they had no evidence that North Korea or any other state was involved. One analyst in Vietnam had earlier said that the attacks originated in the United Kingdom, while the South Koreans insisted that North Korea's telecommunications ministry was behind them. Some people even thought it was all a hoax orchestrated by the South Korean government or activists attempting to incite U.S. action against the North Korean regime.

These attacks were, by most accounts, rather ineffectual and fairly unsophisticated—no data was lost, and the DDoS method is considered a rather blunt instrument—which in part explains why the situation did not escalate. But what happens when more countries can build Stuxnet worms, and even more sophisticated weapons? At what point does a cyber attack become an act of war? And how does a country retaliate when the instigator can almost always cover his tracks? Such questions will have to be answered by policy-makers the world over, and sooner than they expect. Some solutions to these challenges exist, but most options, like international treaties governing cyber attacks, will require substantial investment as well as honest dialogue about what we can and cannot control.

The episodes that prompt these discussions will probably not be state-to-state cyber warfare; a more likely driver will be state-sponsored corporate espionage. States can contain the fallout of attacks on their own governmental networks, but if companies are targeted, the attacks are much more public and can affect more people if user or customer data is involved. Globalization also makes digital corporate espionage a more fruitful endeavor for states. As companies look to expand their reach into new markets, inside information about their operations and future plans can help local entities win contracts and regional favor. To examine why this is true and what it means for the future, we have to look, again, at China.

While China is by no means the only country engaging in cyber attacks on foreign companies, today it is the most sophisticated and prolific. Beijing's willingness to engage in corporate espionage, as well as to sanction its companies to do the same, results in a heightened vulnerability for foreign corporations, not just those looking to work in China but those everywhere in the world. The previously mentioned Chinese cyber attack against Google and dozens of other companies in 2009 is hardly an isolated case; in only the past few years, the industrial-espionage campaign led by Chinese spy agencies has targeted American companies producing everything from semiconductors and motor vehicles to jet-propulsion technology. (Of course, corporate espionage is not a new phenomenon. In one famous nineteenth-century example, England's East India Company hired a Scottish botanist to smuggle Chinese plants and secrets from China into India—which he did successfully, dressed as a Chinese merchant—to break the Chinese monopoly on tea.)

What is new about this latest iteration of corporate espionage is that, in the digital era, so much work can be done remotely and near-anonymously. As we'll see shortly in our discussion of automated warfare, this is a crucial new technological development that will affect many areas in our future world. We live in an age of expansion, and as China and other emerging superpowers seek to expand their economic foothold around the world, digital corporate espionage will greatly enhance their abilities to grow. Whether officially state-sponsored or simply encouraged by the state, hacking into competitors' e-mails and systems to obtain proprietary information will certainly give players an unfair advantage in the market. Several business leaders of major American corporations have told us in confidence about deals they lost in Africa and other emerging markets because of what they believe to be Chinese spying or theft of sensitive information (which was then used to thwart or commandeer their deals).

Today, the majority of cases of corporate espionage between China and the United States appear to involve opportunists rather than the visible hand of the state. There was the Chinese couple in Michigan who stole trade information related to General Motors' research into hybrid cars (which the company estimated to be worth $40 million) and tried to sell it to Chery Automobile, a Chinese competitor. There was the Chinese employee of Valspar Corporation, a leading paint and coatings manufacturer, who illegally downloaded confidential formulas valued at $20 million, intending to sell them to China, and the DuPont chemical researcher who stole information on organic light-emitting diodes, which he planned to give to a Chinese university. None of these actors was tied directly to the Chinese government, and in fact they may simply have been private individuals looking to profit from confidential trade secrets. But we also know that in China, where most major companies are state-owned or heavily influenced by the state, the government has conducted or sanctioned numerous intelligence-gathering cyber attacks against American companies. There can be little doubt that the attacks we know about represent a small percentage of those attempted, whether successful or not.

The United States will not take the same path of digital corporate espionage, as its laws are much stricter (and better enforced) and because illicit competition violates the American sense of fair play. This is a difference in values as much as a legal one—as we discussed earlier, China today does not rate intellectual property rights very highly. But the disparity between American and Chinese firms and their tactics will put both the government and the companies of the United States at a distinct disadvantage. American firms will have to fiercely protect their own information and patrol their network's borders, as well as monitor a range of internal threats (all of the individuals in the above examples legitimately worked for those companies), just to remain competitive.

. . .

The current economic espionage will continue for decades, both between the United States and China and between other nations that gain the required technical capabilities and see the competitive advantages it offers. There will be no dramatic escalation for the same reason that we'll have an ongoing but relatively stable Code War: the lack of attribution in cyber attacks. The Chinese government is free to support or partake in any number of cyber attacks against foreign companies or human-rights organizations so long as their involvement cannot be definitively proven.*

But there are strategies we can use to mitigate the damage caused by cyber attacks in addition to introducing some vulnerability on the part of the attackers. One idea comes from Microsoft's Craig Mundie: virtual quarantine. As we've described, many cyber attacks today come in the form of DDoS attacks and regular denial-of-service (DoS) attacks, which require the use of one "open" or insecure computer on a network that the attacker can use as a base of operations to build his "zombie army" of compromised devices. (DoS attacks could be generated by a small number of hyperactive attacking machines; DDoS attacks are generated by a large, *distributed*—hence the extra "D"—network of attacking machines, often comprised of hacked computers owned by everyday users ignorant of the fact that their computers are being manipulated in this way.) One neglected or unprotected device on the network—a never-used laptop in a science lab, or a personal computer an employee brings to work—can become the attacker's base and then compromise the whole system.†

Quarantine mechanisms contain this attack by enabling the ISP to shut off an infected computer as soon as it recognizes it, unilaterally and without owner authorization, taking the computer off-line. "The basic

---

* Eventually, the Chinese government will be caught red-handed in one of these industrial attacks. If the case is presented to the United Nations Security Council, no resolution will ever be approved, owing to China's veto power, but the outcome will nevertheless be serious geopolitical embarrassment.

† There's an important distinction that needs to be made here. For the purposes of DoS and DDoS attacks, it's not always relevant whether any compromised computers are inside or outside the target's network. Where it matters most is in industrial espionage, when the goal is information extraction; in those cases, computers must be inside the network.

premise is that when you have a network disease, you have to find a way to slow the spread rate," Mundie explained. "We quarantine people involuntarily, but in cyberspace we haven't yet decided that quarantining is the right thing to do." When any machine shows signs of virus or disease, it must be "isolated, contained and healed before being exposed to healthy systems," he added. Users often don't recognize when their computers have been compromised, so allowing the ISPs to conduct these actions will bring about a much faster resolution. Depending on how the mechanism works and what kind of attack is being used, the attackers may or may not recognize that the infected device is off-line—but the user would find his Internet connection inoperable, by mandate of the ISP. By denying the attackers the ability to reach through the infected computer, the harm they can do is greatly reduced.

In Mundie's vision, there would be a neutral international organization to which ISPs could report the IP addresses of infected computers. This way ISPs and states around the world could refuse to let quarantined IP addresses into their online space, cutting off the range of the cyber attack. In the meantime, investigators could watch the cyber attackers from a distance (the attackers would not know the device had been quarantined) and gather information about them to help trace the origin of the attacks. Only when the user had certifiably cleaned his device (with special antivirus software) would his IP address be released from quarantine. In addition to an international organization leading these changes, we might see in parallel the creation of an international treaty around the automatic takedown mechanism. International agreement about swift action to deal with infected networks would be a big step forward in fighting cyber attacks. States that do not agree to the treaty might risk having their whole country considered quarantined, thus putting it off-line for much of the world's users.

Stronger network security will improve the odds for potential targets well before any quarantining is required. One of the basic problems in computer security is that it typically takes much more effort to build defenses than to penetrate them; sometimes programs to secure sensitive information rely on 10 million lines of code while attackers can penetrate them with only 125 lines. Regina Dugan, a senior vice-president at Google, is a former director of DARPA (the Defense Advanced Research Projects Agency), where her mandate included advancing cybersecurity for the U.S. government. She explained to us that, to effectively counter this imbalance, "We went after the technological shifts that would change that basic asymmetry." And, like Mundie, Dugan and DARPA turned to biology as one of the ways to counter the imbalance: They brought together cybersecurity experts and infectious-disease scientists; the result was a program called CRASH, the Clean-Slate Design of Resilient, Adaptive, Secure Hosts.

The philosophy behind CRASH recognized that human bodies are genetically diverse and have immune systems designed to process and adapt to viruses that pass through them, while computers tend to be very similar in their structure, which enables malware to attack large numbers of systems efficiently. "What we observed in cybersecurity," Dugan said, "is that we needed to create the equivalent of an adaptive immune system in computer security architecture." Computers can continue to look and operate in similar ways, but there will have to be unique differences among them developed over time to protect and differentiate each system. "What that means is that an adversary now has to write one hundred and twenty-five lines of code against *millions* of computers—that's how you shift the asymmetry." The lesson learned is undoubtedly applicable beyond cybersecurity; as Dugan put it, "If that initial observation tells you this is a losing proposition, you need something foundationally different, and that in and of itself reveals opportunities." In other words, if you can't win the game, change the rules.

Still, despite some tools for dealing with cyber attacks, lack of attribution online will remain a serious challenge in computer and network security. As a general rule, with enough "anonymizing" layers between one node and another on the Internet, there is no way to trace data packets back to their source. While grappling with these issues, we must remember that the Internet was not built with criminals in mind—it was based on a model of trust. It's challenging to determine who you are dealing with online. Information-technology (IT) security experts get better at protecting users, systems and information every day, but the criminal and anarchic elements on the web grow equally sophisticated. This is a cat-and-mouse game that will play out as long as the Internet exists. The publication of cyber-attack and malware details will help, on a net level; once the components of the Stuxnet worm were unpacked and published, the software it used was patched and cyber-security experts could work on how to protect systems against malware like it. Certain strategies, like universal user registration, might work too, but we have a long way to go before Internet security is effective enough everywhere to prevent simple cyber attacks. We are left once again with the duality of the online world: Anonymity can present opportunities for good or ill, whether the actor is a civilian, a state or a company, and it will ultimately depend on humans how these opportunities manifest themselves in the future.

To summarize: States will long for the days when they only had to think about foreign and domestic policies in the physical world. If it were possible to merely replicate these policies in the virtual realm, perhaps the future of statecraft would not be so complex. But states will have to contend with the fact that governing at home and influencing abroad is far more difficult now. States will pull the most powerful levers they have, which include the control they hold over the Internet in their own countries, changing the online experiences of their citizens and banding together with like-minded allies to exert influence in the virtual world. This disparity between power in the real world and power in the virtual world presents opportunities for some new or underappreciated actors, including small states looking to punch above their weight and would-be states with a lot of courage.

States looking to understand each other's behavior, academics studying international relations, and NGOs and businesses operating on the ground within sovereign territory will need to do separate assessments for the physical and virtual worlds, understanding which events that occur in one world or the other have implications in both, and navigating the contradictions that may exist between a government's physical and virtual foreign and domestic policies. It is hard enough to get this right in a world that is just physical, but in the new digital age error and miscalculation will occur more often. Internationally, the result will be more cyber conflict and new types of physical wars, and, as we will now see, new revolutions.

CHAPTER 4

The Future of Revolution

We all know the story of the Arab Spring, but what we don't know is what comes next. There can be little doubt that the near future will be full of revolutionary movements, as communication technologies enable new connections and generate more room for expression. And it's clear that certain tactical efforts, like mobilizing crowds or disseminating material, will get easier as mobile and Internet penetration rates rise across many countries.

But despite seeing more revolutionary movements, we'll see fewer revolutionary outcomes—fully realized revolutions resulting in dramatic and progressive political turnover. A lack of sustainable leaders combined with savvier state responses will impede profound change (both good and bad) on the scale of the Arab revolutions that began in late 2010. Throughout history, the technologies of the time have stimulated and shaped how revolutions developed, but at a fundamental level all successful revolutions share common factors, like institutional structure, outside support and cultural cohesiveness. The historical record is littered with failed attempts that lacked these basic elements, from Russian revolutionary efforts prior to 1917 through Iraq's Shia uprising in 1991 and the 2009 Green Revolution in Iran. Modern technology, powerful as it is, cannot work miracles, though it can improve the odds of success dramatically.

With so many people connected in so many places, the future will contain the most active, outspoken and globalized civil society the world has ever known. In the beginning of revolutionary movements, the noisy nature of the virtual world will impede the ability of state security to keep up with and crush revolutionary activity, enabling a revolution to start. But how quickly this can happen presents a new problem, since leaders will then have to operate in the physical world of parliaments, constitutions and electoral politics—none of which they'll have the skill or experience to navigate effectively.

Easier to Start . . .

As connectivity spreads and new portions of the world are welcomed into the online fold, revolutions will continually sprout up, more casually and more often than at any other time in history. With new access to virtual space and to its technologies, populations and groups all around the world will seize their moment, addressing long-held grievances or new concerns with tenacity and conviction. Many leading these charges will be young, not just because so many of the countries coming online have incredibly young populations—Ethiopia, Pakistan and the Philippines are three examples where the majority of the population is under the age of thirty-five—but also because the mix of activism and arrogance in young people is universal. They already believe they know how to fix things, so, given the opportunity to take a public stand, they won't hesitate.

Every society in the future, including those that adopted Internet technology early on, will experience different forms of protest in which communication technologies are used to organize, mobilize and engage the international community. The platforms protesters use today—Facebook, Twitter, YouTube and others—will morph into even more constructive vehicles, as developers around the world find new ways to utilize the videos, images and messages related to their particular missions. The world will be introduced to more digital activists, branded heroes by the international community, as they work to become ambassadors for their cause. Countries that have not yet had their first big protest in the new digital age will experience it on a global scale, with the world watching and potentially exaggerating its significance. Democratic societies will see more protests related to perceived social injustice and economic inequality, while people in repressive countries will demonstrate against issues like fraudulent elections, corruption and police brutality. There will be few truly new causes, merely better forms of mobilization and many more participants.

Staging a revolt used to be exclusive to the subset of individuals with the right weapons, international backing and training. Much of this exclusivity has been shattered as communication technologies break down age, gender, socioeconomic and circumstantial barriers that previously prevented individuals from taking part. Citizens will no longer experience injustice in isolation or solitude, and this globalized feedback loop where people all around the world can comment and react will inspire many populations to stand up and make their feelings known. As the revolutions of the Arab Spring demonstrated, once the so-called fear barrier has been broken down and a government appears newly vulnerable, many otherwise obedient or quiet citizens don't hesitate to join in. One of the positive consequences of social media in the Arab revolutions, for example, was that women were able to play a much greater role, given the choice of expressing themselves on social networks when going to the streets was too risky (although many women did take the physical risk). In some countries, people will occasionally organize protests online or in the streets every day, simply because they can. We saw this when we visited Libya in 2012. As we met with ministers in the transitional government in Tripoli, they mentioned casually that there were small groups of protesters nearly every morning. Were they worried?, we asked. Some were, but others shook their heads, almost chuckling, and said it was a natural reaction after more than forty years of oppression.

Virtual space offers new avenues for dissent and participation, as well as new protections for potential revolutionaries. For the most part, dissidents will find their world safer due to the mass adoption of communication technologies, despite the fact that the physical risks they face will not change. (Nor will connectivity shield all activists equally; in countries where the government is very technically capable, dissidents may feel as vulnerable online as they do on the streets.) Arrests, harassment, torture and extrajudicial killings will not disappear, but overall, the anonymity of the Internet and the networked power of communication technologies will provide activists and would-be participants with a new layer of protective insulation that encourages them to continue on.

Certain technological developments will assist activists and dissidents significantly. Accurate real-time translation software enables information-sharing beyond borders. Reliable electronic access to outside information and to diaspora communities helps counter intentionally misleading state narratives and amplifies the size of the support base in a demonstrable way. And secure electronic platforms that facilitate money transfers or information exchange further connect protesters to outside sources of support without compromising their current position.

In these new revolutionary movements, there will be more part-time and anonymous activists than today, simply because citizens will have greater agency over when and how they rebel. Once, being a revolutionary entailed total personal commitment, but today, and even more so in the future, multifaceted technological platforms will allow some to participate full time while others contribute on their lunch breaks. Activists in the future will benefit from the collective knowledge of other activists and people around the world, particularly when it comes to protecting themselves—secure protocols, encryption tools and other forms of electronic security will be more widely available and understood. Most of the people who will come online in the next decade live under autocratic or semi-autocratic governments, and history suggests that theocracies, personality cults and dictatorships are much harder to maintain in an era of expanding information dissemination; one only need recall the contributory role of the glasnost ("openness") policy to the collapse of the Soviet Union. In the end, we'll see a pattern emerge across the world in which populations with access to virtual space and new information will continually protest against their repressive or non-transparent governments online, in effect making the state of revolutionary gestures permanent.

Connectivity will change how we view opposition groups in the future. Tangible organizations and parties will still operate inside countries, but the profusion of new participants in the virtual town square will dramatically reshape the activists' landscape. Most people will not identify themselves with a single cause but instead will join multiple issue-based movements spread over many countries. This trend will both help and frustrate campaign organizers, for it will be easier to estimate and visualize their support network but it will be less clear how interested and committed each participant is. In countries where freedom of assembly is limited or denied, the opportunity to communicate and plan in virtual space will be a godsend, irrespective of who joins in. But generally, it will be up to those in leadership positions to make the strategic decision as to whether their movements actually have the support of the masses, rather than being a very large echo chamber.

For opposition groups, the online world offers new possibilities for critical tasks like fund-raising and branding. Organizations may choose to present themselves differently in different corners of the Internet to reach different demographics. A Central Asian resistance group might downplay its religious overtones and champion its liberal positions while on English-language platforms dominated by Western users, and then do the opposite on the networks within its own region. This is not unlike what the Muslim Brotherhood and other Islamist parties do today, or how Al Jazeera's autonomous English- and Arabic-language operations differ in tone and coverage. (For one example: On a designated day of protest in the early stage of the Syrian uprising in 2011, Al Jazeera English was quick to report on the number of protester deaths but, oddly, the Al Jazeera Arabic website did not, focusing instead on a minor overture by Bashar al-Assad to the country's Kurdish minority. Some analysts suggested that the disparity was due to the Arabic station's political deference to Iran, Syria's ally and a neighbor of Qatar, the home of Al Jazeera.)

While the branding possibilities for these groups grow, the old model of an opposition organization is shifting: Groups today have websites instead of offices; followers and members instead of staff; and they use free and publicly available platforms that liberate them from many fixed costs. There will be so many of these digital fronts in the future that competition for attention between groups around the world will grow fierce.

The profusion of new voices online and the noise they'll generate will require all of us to adjust our definition of a dissident. After all, not everyone who speaks his or her mind online—which to some degree is almost everyone with an Internet connection—can be branded a dissident. The people who surface in the next wave of dissident leaders will be the ones who can command a following and crowd-source their online support, who have demonstrable skill with digital marketing tools, and, critically, who are willing to put themselves physically in harm's way. Digital activism, especially when done remotely or with anonymity, lowers the stakes for would-be protesters, so true leaders will distinguish themselves by taking on physical risks that their virtual supporters cannot or will not. And it's more likely than not that those who have deep knowledge of constitutional reform, institution building and governance issues but lack the tech savvy of other activists will run the risk of being left behind, finding it difficult to stand out in a virtual crowd and to prove their value to new, young leaders (who may fail to understand the true relevance of their experience).

Future revolutionary movements, as we've said, will be more transnational and inclusive than many (but not all) previous revolutions, extending well beyond traditional boundaries of nationality, ethnicity, language, gender and religion. During a trip to Tunisia in 2011, we met with activists from the Jasmine Revolution near the first anniversary of their successful uprising, and when we asked why their revolution set off a chain of others in rapid succession, they acknowledged similar grievances and then pointed to their regional networks. They could build relationships easily with strangers who spoke Arabic and lived in the Middle East, they said, not just because of shared language and culture, but because they often had friends in common. Extensive social connections that already existed were activated and accelerated as revolutionary spirit swept the region, resulting in the exchange of strategies, tools, money and moral support.

But even these large networks had their limit, which was roughly the perimeter of the Arab world. In the future, this won't be true. Sophisticated translation software, which can handle regional accents and is done simultaneously, will enable an Arabic-speaking activist in Morocco to coordinate in real time with an activist in Bangkok who speaks only Thai. Innovative voice translation, streaming gestural interfaces and, eventually, holographic projections will open the floodgates to the formation of much broader virtual networks than anyone has today. There are an untold number of cultural similarities that have never been fully explored because of the difficulty of communication; in a future revolutionary setting, seemingly random connections between distant populations or people will entail knowledge transfer, outsourcing certain types of duties and amplifying the movement's message in a new and unexpected way.

For some, communication technologies will allow them to engage without risk, and to feel the rewards of activism without putting in much effort. It's fairly easy to re-tweet an antigovernment slogan or share a video of violent police brutality from a safe distance, especially when compared with the risks taken by whoever shot the video. People not directly involved in the movement can feel a genuine sense of empowerment by doing something, *anything,* and online platforms offer them a way to chip in and feel valuable, even if what they're doing has little effect on the ground. For people inside a country where there is some risk of being caught by their tech-savvy regime, however, virtual courage does carry risks.

It's certainly possible for a teenager in Chicago or Tokyo to contribute in some significant way to a campaign across the world. After Egypt's external communications capabilities were cut by the Mubarak regime, many observers turned to a Twitter account started by a twenty-something graduate student in Los Angeles for what they perceived to be credible information; the student, John Scott-Railton, posted updates about the protests gleaned from Egyptian sources limited to landline phones. For a time, his @Jan25voices Twitter handle was a major conduit of information about the uprising—this despite his not being a journalist or a fluent Arabic speaker. But while Scott-Railton was able to garner some popular attention for his tweets, there are limits to what someone with his profile could achieve in terms of influencing policy-makers.

Perhaps a more important example is Andy Carvin, who curated one of the most important streams of information in both the Egyptian and Libyan revolutions, with tens of thousands of followers and countless journalists globally who knew that Carvin himself (a senior NPR strategist) had the journalistic standards of a professional reporter and so would tweet or re-tweet only things he could verify. He became a one-man filter of enormous influence, cultivating and vetting sources.

Ultimately, though, however talented the Andy Carvins or John Scott-Railtons of the world are, the hard work of revolutionary movements is done on the ground, by the people inside a country willing to take to the streets. You cannot storm an interior ministry by mobile phone.

The opportunity for virtual courage will shape how protesters themselves operate. Global social-media platforms will give potential activists and dissidents confidence in the belief that they have an audience, whether or not this is true. An organization might overestimate the value of online support, and in doing so neglect its other, more difficult priorities that would actually give it an edge, like persuading regime administrators to defect. The presence of a large virtual network will encourage some groups to take more risks, even if escalation isn't warranted. Full of confidence and courage from the virtual world, a given opposition force will launch campaigns that are immature or ill-advised, the inevitable end result of the breaking down of traditional control mechanisms around revolutionary movements. These trends in virtual courage, for both outsiders and organizers, will have to play out for some time before opposition groups learn how to utilize them effectively.

In all, increased public awareness of revolutions and campaigns around the world will give rise to a culture of revolutionary helpers. There will be a wide range of them: some useful, some distracting and some even dangerous. We'll see smart engineers developing applications and security tools to share with dissidents, and vocal Internet aggregators will use the volume of the crowd to apply pressure and demand attention. No doubt some people will create specialized devices to smuggle into countries with protest movements, handsets that come loaded with encrypted apps that allow users to publish information (texts, photos, videos) without leaving any record on the phone—without a record, a phone contains no evidence of a crime and is thus useless and anonymous to any security thug who finds it.

We'll also see a wave of revolution tourists, people who spend all day crawling the web for online protests to join and help amplify just for the thrill of it. Such actors might help sustain momentum by disseminating content, but they'll be uncontrollable, without filter or oversight, and their narratives might skew expectations for people on the ground taking risks. Finding ways to utilize new participants while exerting quality control and effectively managing expectations will be the key task for effective opposition leaders, who will understand how much else is required for a successful revolution.

. . . But Harder to Finish

The rapid proliferation of revolutionary movements across newly connected societies ultimately will not be as threatening to established governments as some observers predict, because for all that communication technologies can do to transform revolutions in ways that tip the balance in favor of the people, there are critical elements of change that these tools cannot effect. Principal among them is the creation of first-rate leaders, individuals who can keep the opposition intact during tough times, negotiate with a government if it opts for reform, or run for office, win and deliver on what the people want if a dictator flees. Technology has nothing to do with whether an individual has the attributes to fill the role of statesman.

In recent years, we've seen how large numbers of young people, armed with little more than mobile phones, can fuel revolutions that challenge decades of authority and control, hastening a process that has historically taken years. It's now clear how technology platforms can play a prominent role in toppling dictators when used resourcefully. Given the range of outcomes possible—brutal crackdown, regime change, civil war, transition to democracy—it's also clear that it's the people who make or break revolutions, not the tools they use. Traditional components of civil society will become even more important as online crowds swarm the virtual public square, because while some of the newly involved participants (like activist engineers) will be highly relevant and influential, many more, as we've said, will be little more than amplifiers and noise-generators along for the ride.

Future revolutions will produce many celebrities, but this aspect of movement-making will retard the leadership development necessary to finish the job. Technology can help find the people with leadership skills—thinkers, intellectuals and others—but it cannot create them. Popular uprisings can overthrow dictators, but they're successful afterward only if opposition forces have a good plan and can execute it. Otherwise the result is either a reconstitution of the old regime or a transition from a functioning regime to a failed state. Building a Facebook page does not constitute a plan; actual operational skills are what will carry a revolution to a successful conclusion.

The term "leaderless" has been used to describe the Arab Spring, by both observers and participants, but this is not quite accurate. True, in the day-to-day process of demonstrating it's certainly possible to retain a decentralized command structure—safer too, since the regime cannot kill the movement by simply capturing the leaders. But over time, some sort of centralized authority must emerge if the movement is to have any direction. The rebel fighters who faced down Muammar Gadhafi for months were not a coherent army, but by February 27, 2011, within two weeks of the first public protests in Libya, they had formed the National Transitional Council (NTC) in Benghazi. Comprising prominent opposition figures, regime defectors, a former army official, academics, attorneys, politicians and business leaders, the NTC's executive board functioned as an opposition government, negotiating with foreign countries and NATO officials in the fight against Gadhafi. The NTC's chairman, Mahmoud Jibril, served as the country's interim prime minister until late October 2011, shortly after Gadhafi was captured and killed.

In Tunisia, by contrast, the revolution occurred so quickly that there was no time to form an opposition government like the NTC. When President Zine el-Abidine Ben Ali fled, the Tunisian state remained intact. Citizens continued to protest the government until all remaining members of Ben Ali's Constitutional Democratic Rally party resigned and an interim government the masses deemed suitable was in place. Had government officials been less responsive to the population's demands, launching crackdowns instead of reshuffling positions, Tunisia might have followed a very different and less stable path than it did. (Interestingly, many of the leaders elected in the October 2011 Tunisian elections were former political prisoners, who had a different and perhaps more personal level of credibility with the population than returning exiles.) Tunisia's prime minister, Hamadi Jebali—himself a former political prisoner—told us that, in his view, the first post–Ben Ali regime minister of the interior ought to be a "victim of the ministry of the interior." As such, he appointed to this position Ali Laârayedh, who under the previous regime spent fourteen years in prison, mostly in solitary confinement.

The downside of an acceleration in the pace of a movement is that organizations and their ideas, strategies and leaders have a far shorter gestation period. History suggests that opposition movements need time to develop, and that the checks and balances that shape an emergent movement ultimately produce a stronger and more capable one, with leaders who are more in tune with the population they intend to inspire. Consider the African National Congress (ANC) in South Africa. During its decades of exile from the apartheid state, the organization went through multiple iterations, and the men who would go on to become South African presidents (Nelson Mandela, Thabo Mbeki and Jacob Zuma) all had time to build their reputations, credentials and networks while honing their operational skills. Likewise with Lech Walesa and his Solidarity trade union in Poland; a decade passed before Solidarity leaders could contest seats in parliament, and their victory paved the way for the fall of communism.

Most opposition groups spend years organizing, lobbying and cultivating leaders. We asked the former secretary of state Henry Kissinger, who has met with and known almost every major revolutionary leader of the past forty years, what is lost when that timetable is advanced. "It is hard to imagine de Gaulles and Churchills appealing in the world of Facebook," he said. In an age of hyper-connectivity, "I don't see people willing to stand by themselves and to have the confidence to stand up alone." Instead, a kind of "mad consensus" will drive the world and few people will be willing to openly oppose it, which is precisely the kind of risk that a leader must take. "Unique leadership is a human thing, and is not going to be produced by a mass social community," Kissinger said.

Without statesmen and leaders, there won't be enough qualified individuals to take a country forward, running the risk of replacing one form of autocracy with another. "The empowered citizen," Kissinger said, "knows the technique of getting people to the square, but they don't know what to do with them when they are *in* the square. They know even less of what to do with them when they have won." These people can get easily marginalized, he explained, because their strategies lose effectiveness over time. "You can't get people to the square twenty times a year. There is an objective limit, and no clear next phase." And without a clear next phase, a movement is left to run on its own momentum, which inevitably runs out.

There are a number of activists on the street who, while critical of their own revolutions and follow-through, would take issue with Kissinger's view. One such man is Mahmoud Salem, an Egyptian blogger turned activist, who became a spokesperson of sorts for his country's 2011 revolution. Salem is highly critical of his fellow Egyptians for what he saw as an inability to move past the short-term goal of unseating Mubarak and opening the political system to competition; but his critique is one of Egypt, not of the revolutionary model for the new digital age. As he wrote in June 2012, just after Egypt's first post-revolution presidential election, "If you are a revolutionary, show us your capabilities. Start something. Join a party. Build an institution. Solve a real problem. Do something except running around from demonstration to march to sit-in. This is not street work: real street work means moving the street, not moving in the street. Real street work means that the street you live in knows you and trusts you, and will move with you." He exhorted street activists to participate in governance and in reforming the culture of corruption against which they protested. This means wearing seat belts, obeying traffic laws, enrolling in the police academy, running for parliament or holding local officials accountable for their actions.

Tina Rosenberg's book *Join the Club: How Peer Pressure Can Transform the World* is yet another defense of what the crowd can achieve. By looking at the importance of human relationships in defining individual behavior and major social trends, she argues that revolutionaries can channel peer pressure to propel individuals and groups toward more desirable behaviors. Perhaps the most compelling evidence for what she describes as a "social cure" is found in the example of the Serbian activist group Otpor, which played a major role in ending Slobodan Milošević's regime. She describes how the group used playful and flashy street theater, pranks, music, slogans and peaceful civil disobedience to break the culture of fear and helplessness. In cracking down on the group, the regime was revealed to be both brutal and at times foolish, and support for Otpor grew.

But more important than what groups like Otpor represent for the past is the role their leaders can play in the future. As Rosenberg points out in a powerful story of Serbian activists from the past training future activists around the world, successful revolutionaries must develop dual strategies for virtual and physical action. Without both, what's left will be an oversupply of celebrities and coattail riders, and not enough trusted leaders. Historically, a prominent position implied a degree of public trust; with the exception of notorious political types like warlords or machine bosses, the visibility of high-profile leaders corresponded with the size of their support base. But in the future, this equation will be inverted: Prominence will come first and easily, and then a person will need to build tangible support, credentials and experience.

We've seen this already with the self-fulfilling prophecies of "buzzworthy" American presidential candidates. Herman Cain, a relative unknown outside the business world, became highly visible for a period in the 2012 presidential campaign, and he was treated as a serious contender by some despite his political unsuitability for the position—something that revealed itself slowly over weeks, but surely would have been discovered instantly had he been vetted by the party establishment. Political celebrities like Cain will exist in multitudes in future revolutionary movements because flash-in-the-pan charismatic figures who have a strong online presence will rise to the top of the pile most quickly. Without the experience of taking political heat, these revolutionary celebrities are likely to be thin-skinned and will be exposed easily if there is no substance behind their flash.

How opposition movements handle the challenge of finding sustainable leaders will depend on where they are and how many resources they have. In countries where the revolutionary movements are underfunded and under the nose of the regime, pruning the crowds to find genuine leaders will be difficult. In well-resourced and more autonomous movements, however, a crop of consultants might well identify born leaders and subsequently help develop the skills and networks they need. Unlike the run-of-the-mill political consultants of today, these people will have degrees in engineering and cognitive psychology; technical skills; and a much firmer grasp of how to use data to build and fine-tune a political figure. They will take a promising candidate whose prominence exceeds his credentials and measure his political potential through a variety of means: feeding his speeches and writing through complex feature-extraction* and trend-analysis software suites, mapping his brain function to determine how he handles stress or temptation, and employing sophisticated diagnostics to assess the weak parts of his political repertoire.

Many activist groups and organizations will project a virtual front that is far grander than their physical reality. Imagine a new opposi-

---

* Feature extraction automatically identifies the presence, absence or status of important characteristics of a data set. In this case, key features might include the grade level of the writing, the frequency of emotionally charged words and the number of people cited in contexts, thereby indicating mentorship.

tion group being formed just days after a revolution in Algeria, which successfully recruits brilliant digital marketers and designers from the Algerian diaspora in Marseille. The core group consists of only five members, all twenty-somethings barely out of college with almost no prior exposure to politics. Their organization has no track record, but with its sophisticated digital platform they appear to the Algerian public competent, highly motivated and widely networked. In reality, they are disorganized, lacking in vision and wholly unprepared to take on any real responsibility. For groups like this, the dissonance between online presentation and actual operational capability will cause delays and friction within emergent movements. In extreme circumstances, we could see an entire movement that, online, looks like a genuine threat to a regime, when in fact its efforts represent little more than a clever use of technology and actually pose no threat whatsoever. By raising expectations and creating false hope around a movement's prospects for success, opposition groups that can't ultimately rise to meet the challenge may do more harm than good, serving as a costly distraction for the rest of the population.

No doubt every revolution in history has had its share of organizational weaknesses and false prophets, yet in the future, such flaws run the risk of heightening public disenchantment with opposition groups to an extreme degree. If society at large loses faith in a rising movement and its ability to deliver, that's enough to stifle a transformative opportunity. When combined with the instability of leadership, dissonance between the physical and virtual fronts will thoroughly curtail a movement's prospects for support and success in any given country. The consequence of having more citizens informed and connected is that they'll be as critical and discerning about rebels as they are about the government.

This critical eye toward potential opposition forces will have consequences for returning exiles and members of the diaspora, too. Typically, exiles parachute into a country with international support but a limited grasp of the needs and desires of their home population. This disconnection from the realities on the ground has manifested itself in some public flameouts (like onetime Iraqi leader Ahmed Chalabi) and very public struggles (like those of President Hamid Karzai of Afghanistan). On one hand, greater connectivity will decrease the gap between the diaspora communities and the population at home, so returning exiles seeking to have an impact on the revolutionary process will find themselves better suited to connect with local actors. On the other hand, the populations at home will be better informed about the exiles who return (who, no doubt, will have generated long trails of data online about their background and activities), and this information will be used to shape narratives about them before they arrive.

Imagine a prominent Eritrean diaspora member, who made a fortune in the Western media industry, gathering a large virtual constituency with lots of online supporters, both internationally and at home. He might find it difficult to create a *physical* constituency in Eritrea, since many local citizens might be skeptical of his background or his ties to international media. Promises that played well on the international circuit, and with his online audience, might ring hollow to the population back home. Returning to his country expecting to find a path cleared for his political future, he could well watch his promising head start wash away as locals spurn him in favor of a leadership contender they can relate to better.

Successful leaders with ties to the diaspora will be the ones who adopt a sort of hybrid model, whereby the desires of the virtual and physical constituencies are both addressed and somehow reconciled. Winning over and making use of both those groups will be a challenge, but it will be critical for sustainable leadership in the digital age.

A wave of revolutionary false starts will lead successive generations to demand from their opposition groups not only vision but a detailed blueprint of how they intend to build a new country. Such expectations will be true particularly for newer dissident organizations that, in the absence of a track record, still have to demonstrate their bona fides to the public. This follows naturally in the footsteps of technology trends like greater transparency and free access to information. Potential supporters will act more like consumers, less swayed by political ideals than by marketing and product details. There will be more avenues to become a leader (at least in name) and with so many leadership candidates and so little to go on, people will bestow and withdraw their loyalty with ruthless calculation. But competition is as healthy for opposition groups as it is for companies.

Would-be demonstrators looking for a leader will expect any serious opposition group to do its institution-building online, including indicating who the ministers will be, how the security apparatus will be organized, and how goods and services will be delivered. Today, particularly in countries where connectivity is slow to spread, opposition leaders can make vague statements and give assurances that they know what they're doing, but an informed public in the future will demand the details. To the extent that opposition groups exist before a revolution begins—whether in the country itself or in exile—they would be wise to genuinely prepare themselves. Proofs of preparedness to govern will be more than an exercise; the designs will be taken literally as the foundation of a new system. Any opposition group unwilling to produce them or unable to execute them effectively might find lingering praise for its community-organizing skills, but its leadership and governance credentials would certainly be called into question.

Even if an opposition movement presents a credible blueprint, and contains genuine leaders with real skill, there are still a number of uncontrollable variables that could derail a revolution. Tribal, sectarian and ethnic tensions run deep in many societies and remain a minefield for even the most cautious operator to navigate. Internal and external spoilers, like terrorist groups, militias, insurgents and foreign forces, can disrupt the security situation. Many revolutions are spurred by bad economies or fiscal policies, so the slightest economic recalibration (for good or ill) might reverberate through the country and change protesters' minds.

Then there is the dreaded expectations gap. Even if a revolution successfully "finishes," with new players in power and public optimism at its highest point, few new governments will be able to match the expectations and desires of their populations. The consequence of popular uprisings involving many millions more people, thanks in large part to connectivity, is that even more of them will feel abruptly excluded from the political process when the revolution ends.

We saw this directly in Libya and Tunisia when we met with activists and government ministers; neither group felt satisfied or fully appreciated. Following the revolution in Egypt, so many people were unhappy with the way the military rulers, the Supreme Council of the Armed Forces (SCAF), led the country after Mubarak that they subsequently *re*occupied Tahrir Square, the site of the original uprisings, several times. And when the population found itself with limited choices in Egypt's first post-revolution presidential election—Ahmed Shafik, a symbol of the army, and Mohamed Morsi, a symbol of the

Muslim Brotherhood—frustrations and the sense of exclusion only deepened. The degree to which people can feel involved now through connectivity will raise expectations as never before.

New governments will try to meet these demands for accountability and transparency by pursuing "open government" initiatives like publishing ministers' daily schedules, engaging with citizens in online forums and keeping the lines of communication open where possible. Some citizens won't be pacified by anything, however, and in them the ousted political elite will find its own online support network. Clever loyalists will make use of this expectations gap by staying connected to the population online and nurturing its grievances while they attempt to reconstitute the regime. Eventually, they might come to form the new online opposition movement.

Virtual Crackdowns and Containment

Faced with diffuse and omnipresent revolutionary threats, states will look for quick solutions for uprisings that bubble to the surface. They'll have to get creative. Traditional methods like crackdowns and blackouts will become increasingly ineffective as connectivity spreads; the age-old autocratic strategy of suppressing rebellion by violence and rounding up ringleaders is much less relevant in the age of digital protests, online activism and real-time evidence dissemination. Historically, with a few notable exceptions (Tiananmen Square in 1989, the massacre in Hama, Syria, in 1982), crackdowns were rarely captured on film and it was very difficult for images and video to spread outside the country. If the regime controlled all the communications channels, the media and the borders, outside dissemination was nearly impossible.

As soon as mobile devices and the Internet became a feature of rebellion and mass protest, regimes adapted their strategy: They shut down the networks. Initially, this tactic seemed to work for several governments, most notably for the Iranian regime during the 2009 postelection protests when an almost complete shutdown quite effectively curtailed a growing opposition movement. Egyptian president Hosni Mubarak had every reason to believe his virtual crackdown would put a stop to the revolutionary agitation in Tahrir Square less than two years later, but, as the story below illustrates, this strategy had already become counterproductive.

In the early hours of January 28, 2011, anticipating widespread anti-government protests later that day, the Egyptian regime effectively shut down all Internet and mobile connections within the country. "Egypt Leaves the Internet" read the headline of one of the earliest blog posts on the event.* It had blocked access to social-networking sites and BlackBerry Internet service a few days earlier, and with this move, the disconnection was complete.† The country's four main Internet service providers—Link Egypt, Telecom Egypt, Etisalat Misr and Vodafone/Raya—were affected, and mobile-phone service was also suspended by all three telecom operators. The largest of the telecoms, Vodafone Egypt, issued a statement that morning that said, "All mobile operators in Egypt have been instructed to suspend services in selected areas. Under Egyptian legislation the authorities have the right to issue such an order and we are obliged to comply with it."

Given that the Egyptian government already controlled the few physical connections to the outside world—like the fiber-optic cables housed in one building in Cairo—the shutdown was a straightforward matter of closing these portals and contacting the big carriers and contractors with their demands. It was later revealed that the regime made it clear to companies like Vodafone that if they did not comply with the shutdown, the Egyptian government would, through its state-owned company Telecom Egypt, physically cut their service through the telecommunications infrastructure in the country (which would damage Vodafone's ability to operate and take a considerable amount of time to undo). The ISPs and the telecom companies were caught completely off-guard—the government had long been a supporter of the expansion of the Internet and mobile services throughout Egypt—and therefore none had made contingency plans. It was a move unprecedented in recent history; other states had interfered with

---

* The post, by the Internet research firm Renesys, displayed stunning data charts that showed the near-immediate disconnection of Egypt's ISPs from the global network.

† There was one exception to this all-ISP block: Noor Group, which provided service to several prominent institutions like the Egyptian Stock Exchange and the Egyptian Credit Bureau, was left unrestricted until three days later.

their population's Internet services, but none had ever orchestrated such a coordinated and complete disconnection.

The move backfired. As a number of Egyptians and outside observers later noted, it was the shutting down of the network that truly electrified the protest movement because it brought so many more outraged people to the streets. Vodafone's CEO, Vittorio Colao, concurred. "Hitting one hundred percent of the population on something that everyone thinks is essential, and actually taking it out, triggered a much more irritated and negative reaction than what the government expected," he told us. Several Egyptian activists reiterated this, saying, in effect, *I didn't like Mubarak, but this wasn't my fight. But then Mubarak took away my Internet and he made it my fight. So I went to Tahrir Square.* This galvanizing act lent the movement a considerable momentum; had it not occurred, it's possible that events in Egypt would have turned out very differently.

When the regime's request to shut down the network came through, Colao said that Vodafone's first move was to "make sure, from a legal point of view, that we were confronted with a legitimate request. It could be questionable, but it needed to be legal." All telecommunications providers were required to have licensing contracts with the state, so once Vodafone determined that the request was legitimate, it had no choice: "We might not have liked the request, but not honoring it would be a breach of the law."

Soon after, while Internet and mobile-phone service was suspended in Egypt, Vodafone faced another test: The government approached it and other operators in the country to send out its messages over the companies' short-message-service (SMS) platform. This, Colao told us, was where Vodafone played a positive role. At first, he said, the government's tone was procedural: Tonight there will be a curfew from six to nine. "This is one command you can do," Colao explained. The second type of message was patriotic, saying something like, Let's all be friends and love our nation—also fine, said Colao. "But at a point it became incredibly political and one-sided, and that is where you can't ask the local Vodafone staff to say to their own government, *We can't comply with Egyptian law.* We raised the issue with the Egyptian embassy, Hillary Clinton, and the U.K. government, and then Vodafone Group PLC"—the parent company—"put out a statement saying that we [would refuse the government request]. That's what stopped the SMS messages. We were stopped for twenty-four hours for voice calls and four or five days for SMS. SMS is what they considered the threat."

Governments and operators alike will take a lesson from Egypt's failed shutdown tactic. Inside the country, it mobilized masses, and outside, it enraged the international community. Within days of the shutdown, external companies and activists had developed alternative ways for Egyptian citizens to connect again, albeit patchily. A Paris-based nonprofit, French Data Network, opened up Internet access through dial-up connections (available to anyone with an international landline), while Google launched a tweet-by-phone service called Speak2Tweet, which allowed callers to dial one of three numbers and leave a voice mail, which would then be posted as a tweet.

Vittorio Colao told us that after the events in Egypt, major telecom carriers came together to discuss how to prevent such a thing from happening again, and how to take a common position in case it does. Ultimately, he said, "We decided that this has to be discussed within the International Telecommunication Union"—the United Nations special agency for global telecommunications—"to exactly define the rules of engagement." In the future, other governments will surely look to the Egyptian shutdown episode and reevaluate their own odds of survival if they disrupt the connectivity of their populations. Moreover, with peer-to-peer and other connection platforms that operate without a traditional network gaining in popularity, the impact of shutting down communications networks is drastically reduced. Irrational governments, or regimes in a panic, might still consider the extreme step of literally severing the connections at the borders: disconnecting fiber cables, destroying cell towers. But this step would incur such serious economic damage to the country—all financial markets, currency markets and businesses that use external data to operate would fail—that it's very unlikely any regime would take it.

Repressive governments, though, are nothing if not resourceful, and they will find ways to create leverage and exploit loopholes in the face of restive populations and revolutionary challenges. States will develop new methods that are more subtle and insidious. One strategy that many will employ is the if-you-can't-beat-'em-join-'em plan, whereby instead of trying to limit the Internet, they infiltrate it. As we discussed earlier, states stand to gain a significant edge over citizens in the data revolution because of how much of citizens' information they'll have access to. If a government is worried about an uprising, it could ramp up its Internet-monitoring efforts by trawling social-media networks to look for vocal activists; impersonating dissidents to lure in and capture others; hacking into and adding misinformation to prominent mobilization websites; commandeering the webcam on a laptop or tablet to listen to and watch a dissident's actions without his knowledge; and paying close attention to the inflows of money over electronic platforms to identify outside support. Early-stage infiltration might make the difference between a small demonstration and a national rebellion.

Even if the nature of virtual crackdowns changes, however, physical crackdowns will remain a constant in the repressive-state security playbook. Technology is no match for ground-level brutality, as the horrific examples in Syria's multiyear crackdown have shown. Impossible as it seems in the beginning, the international community *can* become desensitized to violent and graphic content, even when the flow of nightmarish images on videos and photographs actually increases over time. All told, for those governments that are still trying to protect their credibility and deny such crimes, brutal crackdowns will become a much riskier endeavor in the digital age. Increased visibility through global online platforms does protect citizens, and this will, we hope, become even more the case as tools like facial-recognition software improve. For an army officer, the knowledge that one well-timed picture from a citizen's handset could identify and shame him internationally—or lead his own government to throw him under the bus—might encourage him to show restraint or even defect. The same could be said for informal civilian militias that engage in violence on behalf of a regime, like the Zimbabwean gangs that fight for Robert Mugabe's ZANU-PF party.

Instead of infiltration (or at least in addition to it), we expect that many states will adopt a strategy we'll call virtual containment. To relieve the pressure of an agitated, informed public, states will calculate that rather than deny services altogether, it's better to crack a window to allow citizens to vent their grievances in public on the Internet— but, more important, only to a certain degree. Regimes in the future will allow some online dissent, whether by reforming the law or simply not prosecuting the speech, but only on their terms, through specific channels they control. After all, giving a Bolivian environmental activist space to complain about the risks of deforestation is unlikely to substantively threaten the strength of the government.

At first glance, the creation of virtual "venting" spaces will seem like a win-win: Citizens will feel a deeper sense of engagement and perhaps a new degree of freedom, while the government will win points for embracing reform (while avoiding or at least stalling an outright rebellion). Perhaps some repressive states will sincerely see the value in reform and offer policy changes without guile. Many won't; not only would the gestures not be genuine (those governments would be uninterested in citizen feedback), but the state would view such spaces as opportunities for intelligence-gathering. Regimes already understand the strategic value of allowing online activity that can lead to arrests. A decade ago, the Egyptian police's vice squad would troll chat rooms and Internet forums with false identities to entrap gay citizens, then lure them to a McDonald's in Cairo to ambush and arrest them.* In 2011, following the Tunisian revolution, several Chinese dissidents responded to an online call for a Chinese version of the protests in front of popular American chains like Starbucks. The mobilization calls spread throughout Chinese social media and microblogs, at which point the police became aware of them. When activists arrived at the prescribed date and time, they were met with an overwhelming police force that arrested many of them. Had the government crushed this online activity immediately after noticing it, the police would not have been able to follow the virtual activity to find the physical dissidents.

As part of their virtual containment strategies, states will undertake a series of transparency gestures, releasing crumbs but withholding the bulk of information they possess. These states will be congratulated for exposing their own institutions and even their own past crimes. Perhaps a government known for its internal corruption will want to appear to turn over a new leaf by publicly disclosing the graft of its judiciary or of a former leader. Or a regime in a single-party state will release some information that is accurate but not particularly damning or useful, like its health ministry's budget statements.

---

* The Egyptian regime was notoriously harsh on its underground gay community; on one infamous occasion, the Cairo vice squad raided a floating nightclub called the Queen Boat and arrested fifty-five men, dozens of whom were convicted of debauchery and sent to prison.

Designated straw men will emerge to take responsibility and bear the brunt of public anger, and the regime will survive intact. Manufacturing transparent-looking documents and records will not be difficult for these regimes—in the absence of contradictory information (such as leaked original documents), there's little hope of proving them false.

The real challenge for states that adopt the virtual containment approach will be distinguishing between public venting and real opposition online. Computer engineers use the term "noise" to describe data that can be very loud but does not convey a useful signal. Authoritarian governments will encounter a political version of this as they begin to allow freer online discussion. In open societies, laws regarding freedom of speech and hate speech largely define the boundaries for citizens, but in closed countries that lack legal precedents for allowable speech, the government is operating somewhat blindly. It will be very difficult for states to determine the intent behind people's words online—if they're not known dissidents, have no ties to opposition groups and don't stick out in any particular way, how does a government newly committed to open dialogue respond without going too far? This unknowable quality will make digital noise the big wild card for authorities as they struggle to first assess and then react. Getting it wrong, by overreaction or underreaction, could be lethal for a regime. Neglect of an online swell could turn it into an off-line storm, and harshly cracking down on online banter could give a nascent online movement with no real momentum something to rally around.

There are a number of present-day examples of state overreaction to online content, though none have yet resulted in revolution. Two examples from Saudi Arabia in 2011 stand out, and they suggest a model for the escalation path we will see in the future. The first involved a group of conservative clerics who, angered by the Saudi king's decision to grant women the right to vote in the 2015 municipal elections, immediately retaliated against a group of women who had participated in a Women2Drive Campaign (during which several women openly defied Saudi law and got behind the wheel). The clerics decided to make an example of one of the women and sentenced her to ten lashes. As news of her sentence spread, ordinary Saudis took to the Internet to protest and stand up for her, sharing the news far beyond the country's borders. The virtual retaliation of hundreds of thousands of people both in and outside of Saudi Arabia led the government to revoke the decision less than twenty-four hours later. In this instance, the Saudi king's quick reaction stemmed a rising tide, but his very responsiveness suggests a genuine state concern about the threat posed by clamorous online mobs.

The second example comes from a decision to ban a satirical short film about Saudi Arabia's expensive housing market. As with most officially prohibited material throughout history, there is no surer way to drive public interest and demand than by government ban, and this case is no different. The film, *Monopoly*, appeared on YouTube within an hour of the ban, and in just a few weeks had accumulated more than a million views. If the flogging story highlights the importance of swift action to reverse mistakes, this one speaks to the importance of regimes' picking the right battles. They will never be able to predict the trigger that transforms online venting into street protest, so every decision to react or ignore is a gamble. Saudi Arabia has not seen large-scale public protests to date, but as a country with one of the most active social-media populations in the region (with one of the highest rates of YouTube playbacks of any country in the world, no less), it will surely encounter more small battles like those described above, and a miscalculation on any one of them could lead to a much larger problem.

No More "Springs"

As more societies come online, people will look for signs of regional revolutionary epidemics. Some argue that Latin America will be next, because of its serious economic disparities, weak governments, aging leaders and large populations that speak the same language. Others make the case for Africa, where state fragility is the highest in the world, while mobile-phone adoption is skyrocketing and creating the fastest growing mobile market anywhere. Or perhaps it would be Asia, which has the largest number of people living under autocratic rule, runaway economic growth and myriad widespread social, economic and political tensions. There have already been nascent attempts to organize mass protests and demonstrations in Vietnam, Thailand, Malaysia and Singapore, and surely this will continue to build with time.

But even though these regions are becoming more connected and their populations are increasingly exposed to events and the shared grievances of other nationalities, we don't yet have evidence that there will be another iteration of the contagion effect the world saw in the Arab Spring. (It is worth noting, though, that a contagion of protests and demonstrations *will* be easier to achieve, as illustrated by the September 2012 reactions to the infamous video *Innocence of Muslims* in several dozen countries throughout the world.) The Arab world has a unique regional identity not shared by other regions, which has been solidified by historical attempts at unification and pan-Arab sentiments over the decades. And, of course, shared language, culture and similar political systems contributed. As we said earlier, modern communication technologies did not invent the networks that activists and protesters in the Middle East made use of—they amplified them.

In addition, there were established religious networks, which, in the absence of a strong civil society under autocratic rule, were by default the most organized and often most beneficial nongovernmental entity for citizens. All of the Arab leaders who lost power in this wave of revolutions—Tunisia's Ben Ali, Egypt's Mubarak, Libya's Gadhafi and Ali Abdullah Saleh in Yemen—built and operated political systems that stifled the development of institutions, so religious houses and organizations often filled that void (in doing so, they earned the enmity of these dictators; the most prominent groups, like the Muslim Brotherhood in Egypt and the Islamist Ennahda party in Tunisia, were either banned outright or mercilessly persecuted by the state because they constituted such a threat). Over the course of the recent revolutions, mosques became gathering points, imams and other clerics lent legitimacy to the protesters' cause in some cases, and religious solidarity for many people was an important motivation for mobilization.

In other regions, these components are missing. Africa, Latin America and Asia are far too heterogeneous and diverse in culture, language, religion and economics to mirror the Arab model. Regional identity does not exist to the extent that it does in the Middle East, and social, business and political networks are more localized.

However, it's impossible not to see changes on the horizon in all of these regions. They might be country-specific and include a broader range of outcomes than regime change, but nonetheless they will be profound on a political and psychological level. Every country in the world will experience more revolutionary triggers, but most states will weather the storm, not least because they will have the opportunity to watch and learn from other countries' mistakes. A collection of best practices will emerge among states to deflect, diffuse and respond to the charges presented by newly connected publics. (This is a reasonable assumption since the interior ministers in repressive states, responsible for policing and national security, visit with each other to share knowledge and techniques.) Issues like income inequality, unemployment, high food prices and police brutality exist everywhere, and governments will have to make preemptive adjustments to their policies and messages to address public demand more responsively than in earlier times. Even in comparably stable societies, leaders are feeling the pressure of a connected citizenry and recognizing the need for reform or adaptation in the new digital age because no government is invulnerable to these looming threats.

Nobody understands this combination of political pressures and technological challenges better than Singapore's prime minister, Lee Hsien Loong, who is both a regional leader and a computer scientist by training. "The Internet is good for letting off steam," he told us, "but it can also be used to create new fires. The danger we face in the future is that it will be far easier to be against something than for it." Young people everywhere, he explained, always want to be part of something cool, and "this social experience of being against authority means young people no longer need a plan. It has become far too easy for very minor events to escalate into lots of online activity that is exploited by opposition groups."

Lee pointed to a recent event in his own country, known colloquially as "Currygate." "A Chinese immigrant and a Singaporean of Indian descent quarreled over the right to cook curry, given that the aroma seeps through the walls," Lee said. The Chinese man considered his neighbor's constant curry cooking inconsiderate, and, "in typical Singaporean fashion," the two brought in a mediator to resolve the dispute. An agreement was reached: The Indian would cook curry only when his neighbor was out of town. That was the end of it until, years later, the mediator went public with his story. The Indian community in Singapore was outraged, incensed by the idea that the Chinese could dictate when people did or did not cook curry, and the situation escalated quickly. According to Lee, "What began as the declaration of a national curry-cooking day led to thousands of 'likes' and posts and a viral movement that captured the attention of the entire country." Luckily for Lee, the online agitation around curry didn't lead to massive protests in the streets, even though the rhetoric was highly charged at the time.

The protests in Singapore had little to do with curry and everything to do with the growing concerns about foreigners (particularly mainland Chinese) coming in and taking jobs. Unsurprisingly, opposition groups keen to push this agenda found Currygate an easy episode to exploit. For a country like Singapore, which prides itself on stability, efficiency and the rule of law, the broadcasting of such anger from so many citizens revealed a vulnerability in its system: Even in as tightly controlled a space as Singapore, government restrictions and social codes have limited leverage in the online world. For Lee, the episode foreshadowed a tide of online expression that the Singaporean leadership acknowledges will be impossible to roll back. If even the authorities in Singapore are feeling the heat of a newly connected civil society, imagine how nervous more fragile governments in other parts of the world must feel.

We asked Lee how he thought China would handle this transition, given that, in a decade, almost a billion Chinese citizens will become connected in a heavily censored society. "What happens in China is beyond anyone's full control, even the Chinese government," he said. "China will have a difficult time accommodating all of these new voices, and the transition from a minority of the population online to the majority is going to be difficult for the leaders." Concerning the subject of leadership, he added, "Successive generations of Chinese leaders will not have the charisma or communications skills to generate momentum among the population. In this sense, the virtual world will become far cooler and far more relevant to the Chinese people than the physical world." Change, he said, would not just come from people outside the system: "It is people inside the system, the cadres of the Chinese establishment, who are influenced by the [street] chatter and who also have skeptical views of the legitimacy of the government."

We agree with Lee and other regional experts that China's future will not necessarily be bright. Some interpret projections of declining economic growth, an aging population and technology-driven change as indications that the Chinese state will soon be fighting for survival in its current form, while others suggest instead that these impending challenges will ultimately spur even more innovation and problem-solving from China. But ultimately it is difficult for us to imagine how a closed system with 1.3 billion people, huge socioeconomic challenges, internal ethnic issues and robust censorship will survive the transition to the new digital age in its current form. With greater connectivity will come greater expectations, demands and accountability that even the world's largest surveillance state will not be able to control fully. In instances where law enforcement goes too far or cronies of the regime engage in reckless behavior that causes physical harm to Chinese citizens, we will see more public movements demanding accountability. Because ministers loathe embarrassment, pressure from *weibos* and other online forums can result in more pressure and change, eventually curbing the excesses of one-party rule.

So while the Internet may not democratize China overnight, increased public accountability will put at least some pressure on the regime to act on the public's demands for justice. And if economic growth should noticeably slow down, it could create a revolutionary opening for some elements of the population. China will experience some kind of revolution in the coming decades, but how widespread and effective it is will come down to the willingness of the population to take risks both online and in the streets.

Future revolutions, wherever they happen and whatever form they take, may change regimes, but they will not necessarily produce democratic outcomes. As Henry Kissinger told us, "The history of revolutions is a confluence of resentment that reaches an explosive point and it then sweeps away the existing structure. After that, there is either chaos or a restoration of authority which varies in inverse proportion to the destruction of previous authority." In other words, following a successful revolution, "the more authority is destroyed, the more absolute the authority that follows is," Kissinger said. Having experienced successful and failed revolutions over more than forty years, he has deep knowledge of their designs and character. The United States and Eastern Europe are the only cases, according to him, in which the destruction of the existing structure led to the creation of a genuine democracy. "In Eastern Europe," he explained, "the revolutions succeeded because the experience of dictatorship was so bad, and there was a record of being Western and part of the democratic tradition, even if they were never democracies."

While Kissinger's point about the distinctness of Eastern Europe is well taken, we cannot dismiss the role that incentives play in the success of revolutions. We would be remiss to leave out the incredibly important incentive of being able to join the European Union (E.U.). If E.U. membership had not been available as a political motive for liberal elites and populations as a whole and also as a stabilizing factor, we would likely have seen much more backsliding and counterrevolution in a number of different countries. This is why the Western powers had to expand NATO *and* offer E.U. membership.

The absence of this democratic culture is part of the reason the overthrow of dictatorships during the Arab Spring produced, in the eyes of some, merely watered-down versions of autocracies instead of pure Jeffersonian democracies. "Instead of having all power consolidated under one dictator," Kissinger said, "they split themselves into various parties—secular and non-secular—but ultimately find themselves dominated by one Muslim party running a token coalition government." The result will be coalition governments, which *"The New York Times* will welcome as an expression of great democracy," he joked, but really, "at the end of that process stands a government without opposition, even if it comes into being in a one-off election."

Autocratic-leaning coalition governments, Kissinger predicts, will often be the form new governments produced by digital revolutions assume in the coming decades, less because of technology than because of the lack of strong, singular leaders. Without a dominant leader and vision, power-sharing governments emerge as the most viable option to pacify most participants, yet they'll always run the risk of not distancing themselves sufficiently from the previous regime or the older generation of political actors.

Revolutions are but one manifestation of discontent. They stick out in our memories because they can often adopt romantic overtones, and be easily woven into human narratives about freedom, liberty and self-determination. With more technology come more anecdotes that capture our imagination and make nice headlines. Even when unsuccessful, revolutionaries occupy a particular position in our collective history that confers a certain respect, if begrudgingly so. These are highly important components in human political development, central to our understanding of citizenship and social contracts, and the next generation of technologies will not change this.

But while revolutions are how some pursue change within the system or express their discontent with the status quo, there will always be people and groups who pursue the same objectives through the most devastating and violent means. Terrorists and violent extremists will be as much a part of our future as they are our present. The next chapter will delve into the radicalization hotbeds of our future—both in the physical world and online—and explain how an extended battlefield will change the nature of terrorism and what tools we have to fight it.

CHAPTER 5

The Future of Terrorism

As we've made clear, technology is an equal-opportunity enabler, providing powerful tools for people to use for their own ends—sometimes wonderfully constructive ends, but sometimes unimaginably destructive ones. The unavoidable truth is that connectivity benefits terrorists and violent extremists too; as it spreads, so will the risks. Future terrorist activity will include physical and virtual aspects, from recruitment to implementation. Terrorist groups will continue to kill thousands annually, by bombs or other means. This is all very bad news for the broader public, states that already have enough trouble protecting their homeland in the physical world, and companies that will be increasingly vulnerable.

And of course there remains the terrifying possibility that one of these groups will acquire a nuclear, chemical or biological weapon. Because of the developed world's increasing dependence on its own connectedness—nearly every system we have is tied to a virtual network in some way—we're acutely vulnerable to cyber terrorism in its various forms. That applies, of course, even in less-connected places, where the majority of terrorist attacks occur today. The technical skills of violent extremists will grow as they develop strategies for recruitment, training and execution in the virtual world, with the full understanding that their attacks will be more visible than ever before thanks to the increasing reach of global social-media networks.

But despite those gains, communication technologies also make terrorists far more vulnerable than they are today. For all the advantages that living in the virtual world give terrorists (small cells all over the world, destructive activities that are harder to trace), they still have to live physically (eat, be sheltered, be in a physical space from which they use their phones and computers), and that's precisely what makes them more vulnerable in the new digital age. Here we will explore how terrorists will split their time between the physical and virtual worlds and why despite some advantages gained, they will ultimately make more mistakes and implicate more people, making their violent business far more difficult.

New Reach, New Risks

That the Internet provides dangerous information for potential criminals and extremists is well known; less understood is how this access will evolve on a global scale in the future. Many of the populations coming online in the next decade are very young and live in restive areas, with limited economic opportunities and long histories of internal and external strife. It follows, then, that in some places, the advent of the new digital age will also mean an increase in violent activity fueled by the greater availability of technology. A strong indication that that process is under way will be the proliferation of sophisticated homemade explosive devices.

While traveling in Iraq in 2009, we were struck by the notion that it was far too easy to be a terrorist. An Army captain told us that one of the greatest shared fears among American troops on patrol was the hidden roadside IED (improvised explosive device). In the war's early days, IEDs were expensive to produce and required special materials, but with time, bomb-making tools and accessible instructions were widely available to any potential insurgent. The IED of 2009 was cheaper and more innovative, designed to evade now-understood countermeasures with simple adaptations. A bomb with its trigger taped to a mobile phone set on "vibrate" could be detonated remotely by calling that number. (The Americans soon responded to this tactic by introducing jamming systems to cut off mobile communication, with limited success.) What was once a sophisticated and lucrative violent activity (earning insurgents thousands of dollars) had become routine, an option for anyone with a bit of initiative willing to be paid in cigarettes.

If an insurgent's mobile-phone-triggered IED is now the equivalent of a high school science project, what does that tell us about the future? These "projects" are an unfortunate consequence of what the Android creator Andy Rubin describes as the "maker phenomenon" in technology, which outside the terrorism context is often applauded. "Citizens will more easily become their own manufacturers by piecing together versions of today's products to make something that had previously been too hard for an ordinary citizen to build," Rubin told us. The emerging "maker culture" around the world is producing an untold number of ingenious creations today—3-D printers are just the beginning—but as with most technology movements, there is a darker side to innovation.

The future homemade terror device will likely be a combination of "everyman" drones and mobile IEDs. Such drones could be purchased online or at a toy store; indeed, simple remote-control helicopters are already available. The AR.Drone quadricopter, built by Parrot, was one of the top-selling toys of the 2011–2012 Christmas season. These toys are already equipped with a camera and can be piloted by a smart phone. Imagine a more complicated version that uses a Wi-Fi connection it generates itself and that is fitted with a homemade bomb on its undercarriage, producing a whole new level of domestic terror that is just around the corner. The knowledge, resources and technical skills necessary to produce such a drone will certainly be available practically everywhere in the near future. The autonomous navigation capability we discussed previously will become generally available and embeddable on a chip, which will make it easier for terrorists and criminals to stage a drone-based attack without intervention. Improved destructive capacity in physical attacks is just one way the spread of technology will affect global terrorism. Cyber terrorism, of course, is another— the term itself dates back to the 1980s—and the threat will grow only graver. For our purposes, we'll define cyber terrorism as politically or ideologically motivated attacks on information, user data or computer systems intended to result in violent outcomes. (There is some overlap in tactics between cyber terrorism and criminal hacking, but generally the motivations distinguish the two.)

It's hard to imagine extremist groups operating out of caves in Tora Bora constituting a cyber threat, but as connectivity spreads throughout the world, even remote places will have reasonable network access and sophisticated mobile handsets. We have to assume that these groups will also acquire the technical skills necessary to launch cyber attacks. Those changes, and the fact that our own connectedness presents an endless number of potential targets for extremists, are not promising developments.

Consider some straightforward possibilities. If cyber terrorists successfully compromise the network security of a large bank, all of its customers' data *and* money will be at risk. (Even calling in a threat, in the proper circumstances, could cause a run on the bank.) If cyber terrorists target a city's transportation system, police data, stock market or electricity grid, they could bring the daily mechanics of the city to a halt.

The security shields of some institutions and cities will prevent this, but not everyone will have such protection. Nigeria, which struggles with domestic terrorism and weak institutions, is already a world leader in online scams. As the connectivity of the cities of Lagos and Abuja extends to the more restive and rural north (where violent extremism is most prevalent), many would-be scammers could easily be attracted to the cause of a violent Islamist group like Boko Haram (Nigeria's version of the Taliban). Only a handful of new recruits could transform Boko Haram from West Africa's most dangerous terrorist organization into its most powerful cyber-terrorist one.

Cyber-terrorist attacks need not be limited to system interference, either. Narco-terrorists, cartels and criminals in Latin America lead the world in kidnappings, but in the future, traditional kidnapping will be riskier, given trends like precision geo-location in mobile phones. (Even if kidnappers destroy a captive's phone, its last known location will have been recorded somewhere in the cloud. Security-conscious individuals in countries where kidnapping is widespread will likely also have some form of wearable technology, something the size of a pin, which would continuously transmit their location in real time. And some who are most at risk may even have variations of those physical augmentations we wrote about earlier.) Virtual kidnappings, on the other hand—stealing the online identities of wealthy people, anything from their bank details to public social-network profiles, and ransoming the information for real money—will be common. Rather than keep and maintain captives in the jungle, guerrillas in the FARC or similar groups will prefer the reduced risk and responsibility of virtual hostages.

There are clear advantages to cyber attacks for extremist groups: little to no risk of personal bodily harm, minimal resource commitment, and opportunities to inflict a massive amount of damage. These attacks will be incredibly disorienting for their victims, due to the difficulty of tracing the origins of virtual attacks,* as we noted earlier, and they will induce fear among the enormous pool of potential victims (which includes nearly everyone whose world relies on being connected). We believe terrorists will increasingly shift their operations into the virtual space, in combination with physical-world attacks. While the dominant fear will remain weapons of mass destruction (the porousness of borders making it far too easy to smuggle a suitcase-sized bomb into a country), a future 9/11 might not involve coordinated bombings or hijackings, but coordinated physical and virtual-world attacks of catastrophic proportions, each designed to exploit specific weaknesses in our systems.

An attack on America could begin with a diversion on the virtual side, perhaps a large-scale hacking into the air-traffic-control system that would direct a large number of planes to fly at incorrect altitudes or on collision paths. As panic sets in, another cyber attack could bring down the communication capabilities of many airport control towers, turning all attention to the skies and compounding the fear that this is the "big one" we've been fearing. Meanwhile, the real attack could then come from the ground—three powerful bombs, smuggled in through Canada, that detonate simultaneously in New York, Chicago and San Francisco. The rest of the country would watch as the first responders scrambled to react and assess damage, but a subsequent barrage of cyber attacks could cripple the police, the fire department

---

* Cyber attackers cover their tracks by routing data through intermediary computers between themselves and their victims. Such "proxy" computers—which could include hacked computers in homes or businesses around the globe—appear to victims and outsiders as the sources of the attack, and it can be quite challenging to trace through many intermediary layers back to the true sources of cyber attacks. Making matters worse, an attacker can launch a Tor router on the compromised host, spewing obfuscating traffic throughout the compromised network and masking the attacker's intentional activities.

and emergency-information systems in those cities. If that's not terrifying enough, while urban emergency efforts slow to a crawl amid massive physical destruction and loss of life, a sophisticated computer virus could attack the industrial control systems around the country that maintain critical infrastructure like water, power and oil and gas pipelines. Commandeering these systems, called supervisory control and data acquisition (SCADA) systems, would enable terrorists to do all manner of things: shut down power grids, reverse waste-water treatment plants, disable the heat-monitoring systems at nuclear power plants. (When the Stuxnet worm attacked Iranian nuclear facilities in 2012, it operated by compromising the industrial control processes in nuclear centrifuge operations.) Rest assured that it would be incredibly, almost unthinkably difficult to pull off this level of attack—commandeering one SCADA system alone would require detailed knowledge of the internal architecture, months of coding and precision timing. But some kind of coordinated physical and cyber attack is inevitable.

Few terror groups will possess the level of skill or the determination to carry out attacks on this scale in the coming decades. Indeed, because of the vulnerabilities that technology introduces for them, there will be fewer terrorist masterminds altogether. But those that do exist will be even more dangerous. What gives terror groups in the future an edge may not be their members' willingness to die for the cause; it might be how good their command of technology is.

Various platforms will aid extremist groups in planning, mobilization, execution and, more important, as we've already pointed out, recruitment. There may not be many caves online, but those blind spots where all manner of nefarious dealings occur, including child pornography and terrorist chat rooms, will continue to exist in the virtual world. Looking ahead, future terror groups will develop their own sophisticated and secure social platforms, which could ultimately serve as digital training camps as well. These sites will expand their reach to potential new recruits, enable information-sharing among disparate cells and serve as an online community for like-minded individuals. These virtual safe houses will be invaluable to extremists, provided that there are no double agents and that the digital encryption is strong enough. Antiterrorism units, law enforcement and independent activists will try to shut down or infiltrate these sites but will be unable to.

It's just too easy to relocate or change the encryption keys in boundless virtual space and keep the platform alive.

Media savvy will be among the most important attributes for future transnational terrorists; recruitment, among other things, will rely on it. Most terrorist organizations have already dipped a toe into the media marketing business, and what once seemed farcical—al-Qaeda's website heavy with special effects, Somalia's al-Shabaab insurgent group on Twitter—has given way to a strange new reality. The infamous case of Anwar al-Awlaki, the late American-born extremist cleric affiliated with al-Qaeda in Yemen, provides a compelling example. His high profile was largely a result of his own self-promotion—he used viral videos and social networks to disseminate his charismatic sermons internationally. As the first major terrorist YouTube sensation, Awlaki's influence is undeniable—several successful and would-be terrorists cited him as an inspiration—and his prominence earned him a spot on the U.S. government's list of high-value targets. He was killed by a drone strike in September 2011.

Awlaki's social media mastery impressed the billionaire investor and reformist Saudi prince Alwaleed bin Talal al-Saud, who sees this as part of a broad trend across the region. "Even the most anti-Western religious figures in Saudi Arabia are now almost all using technology," he told us, adding that "a number of them are even using mobile devices and increasingly social networks to issue *fatwas*"—Islamic edicts. As Middle East observers know, this is a profound change, particularly in Saudi Arabia, where the clerical establishment is notoriously slow to accept technology. The trend will only continue.

Given the importance of digital marketing for future terrorists, we anticipate that they will increasingly look to infiltrate mobile and Internet companies. Some Islamist groups have already tried to do this. Maajid Nawaz, a former leader in Hizb ut-Tahrir (HT)—a global extremist group that seeks the overthrow of Muslim-majority governments through military coups and the creation of a worldwide Islamist superstate—told us his organization had a policy of recruiting from mobile-phone companies. "We pitched propaganda stalls outside the Motorola offices in Pakistan, then we recruited some Motorola staff, who proceeded to leak the numbers of Pakistan's national newspaper editors," he said. Members of HT would bombard these editors with text messages full of propaganda, talking points and even threats. The recruited Motorola staff further helped HT, according to Nawaz, by concealing its members' identities when they signed up for phone service, allowing them to operate undetected.

If extremist groups don't target the mobile companies themselves, they will find other ways to wield influence on these powerful platforms. Groups like Hamas and Hezbollah tend to gain community support by providing services that the state is unwilling or unable to deliver adequately. Services, support and entertainment all serve to strengthen the credibility of the group and the loyalty of its base. Hamas could develop a family of apps for the cheap smart phones everyone uses, offering everything from health-care information to mobile money exchanges to games for children. This infinitely valuable platform would be built and serviced by Hamas members and sympathizers. Even if the Apple store blocked their applications under order of the U.S. government, or the U.N. took similar action, it would be possible to build apps without any official tie to Hamas and then promote them through word of mouth. The impact this could have on a young generation would be immense.

As global connectivity renders extremist groups more dangerous and more capable, traditional solutions will appear increasingly ineffective. In many parts of the world, simply imprisoning terrorists will have little effect on their network or their ability to influence it. Smuggled handsets will enable extremists to run command-and-control centers from inside prison walls, and the task of confiscating or otherwise limiting the power of these devices will only get harder as the basic components of smart phones—the processors, SIM cards (memory cards used in mobile phones that can carry data from one phone to another) and the rest—get smaller and more powerful.

Such practices have already begun, sometimes in farcical fashion. In 2011, Colombian prison officials stopped an eleven-year-old girl en route to visiting an incarcerated relative in Medellín because of the odd shape of her sweater; they found seventy-four mobile phones and a revolver taped to her back. In Brazil, inmates trained carrier pigeons to fly in phone components, and at least one local gang hired a teenager to launch phones over the prison walls with a bow and arrow. (The boy was caught when one of his arrows struck an officer.)

This is not just taking place in the developing world. A former member of a South Central Los Angeles gang told us that the going rate for a contraband smart phone hovers around $1,000 in American prisons today. Even tablets can be obtained for the right price. He further described how these devices enable well-connected inmates to maintain their illicit business ties from behind bars through popular social-network platforms. In 2010, when inmates in at least six prisons in the U.S. state of Georgia simultaneously went on strike to protest their conditions, their protest was organized almost entirely through a network of illicit mobile phones.

The most compelling (and successful) example of prison activities comes from Afghanistan, a country with one of the lowest rates of connectivity in the world. The Pul-e-Charkhi prison on the outskirts of Kabul is the country's largest prison and among its most notorious. Commissioned in the 1970s and completed during the Soviet occupation, in its initial years tens of thousands of political prisoners were killed there annually and many more were tortured for anti-Communist sentiments. The prison earned a new distinction during the American occupation as a terrorist nerve center. Following a violent riot in 2008 in the prison's Cell Block Three, Afghan authorities discovered a fully operational terror cell—in both senses of the word—that had been used by inmates to coordinate deadly attacks outside the prison walls. The back door to the cell block was covered in live electrical wires, woven through the bars like vines and emitting a soft red glow in the corridor, and the walls were painted with swords and verses from the Koran. Cell Block Three had been taken over by its Taliban and al-Qaeda inmates years earlier, and through a combination of effective smuggling of phones and radios, savvy recruitment within the prison population and threats to the guards and their families, these radicalized inmates had transformed their environment into a prison without walls—a secure perch (safe from aerial drones and other dangers) from which they could expand their organization, run extortion schemes and coordinate terrorist attacks in a city twenty miles away. They recruited petty thieves, heroin addicts and Christians (inmates whose pariah status in Afghan society made them ripe for radicalization) with money or the threat of violence.

After the 2008 riot, relocation of these inmates to different cell blocks was thought to have ended their terror network, or at least severely curtailed its functionality. Yet two years later, following a string of attacks in Kabul, prison officials admitted publicly that the terror cells had re-formed within Pul-e-Charkhi almost immediately, and authorities' attempts to limit their operational capacity by sporadic jamming (to render their contraband mobile phones useless) had largely failed. Pul-e-Charkhi housed many of Afghanistan's high-value inmates, and it was run by the Afghan military with American advisors, yet no one seemed capable of controlling the mobile networks. When Jared accompanied the late special envoy to Afghanistan Richard Holbrooke on a trip to Kabul, he visited the prison and met with one of the incarcerated former ringleaders of Cell Block Three, an extremist leader named Mullah Akbar Agie, to assess how conditions in the prison had changed after the post-riot crackdown. Agie responded to a joking request for his phone number by reaching into his robe and pulling out a late-model feature phone. He proudly jotted down his name and phone number on a slip of paper: 070-703-1073.

The experience at Pul-e-Charkhi suggests the danger of mixing gangs, religious extremists, drug traffickers and criminals in the prisons during the digital age. Outside prison walls, these different networks at times overlap and use the same technological platforms, but when they are put in close proximity inside prisons, with the help of contraband devices they can become dangerous, united nodes. A band of narco-traffickers from a Mexican cartel might share valuable information about cross-border weapon-smuggling networks with an Islamist extremist in exchange for money or a foothold in a new market for the cartel. When both parties reach a mutually beneficial arrangement, each could use his mobile phone to inform his organization of the new collaboration. Deals struck in prison and then followed through in open society will be difficult to intercept, because short of placing all inmates in isolation cells (unrealistic) or shutting down the mobile contraband trade (equally unlikely, despite enormous effort), prison authorities will have limited success in preventing cases like these from materializing.

So if we take it as a given that prison contraband networks will generally outsmart the officials charged with shutting them down, and that mobile phones will remain in high demand for inmates, what options remain to thwart the Pul-e-Charkhi scenario from playing out elsewhere? The most obvious solution is simply cutting off access, jamming the networks so that inmates' illicit phones become little more than expensive platforms for playing Tetris. But it stands to reason that someone could figure out a way to get over that hurdle. Perhaps live pigeons won't work, but small drones designed to look like pigeons and act as mobile Wi-Fi hot spots might.

Monitoring and tapping the mobile activity among prisoners is another option for law enforcement. The intelligence gathered from listening in could, among other things, shed light on how illicit networks operate. A more subversive solution could be to intentionally co-opt the contraband networks by getting devices into prisoners' hands that are actually filled with traps to inadvertently give up information. Loaded with malware that will allow activity on each phone to be traced, these phones would be designed to give up secrets easily without inmates' knowledge. This may ultimately prove more effective than human informants, and safer, too.

Some societies will ensure that a prisoner disappears from the Internet entirely while behind bars. By court order his virtual identity would be frozen, laws would prevent anyone from trying to contact, interact with or even advertise to his frozen profile, and once he was released, he would be required to provide his probation officer with access rights to his online accounts. The digital-age equivalent of an ankle bracelet will be government-imposed software that tracks and restricts online activity, not just for the obvious cases like child molesters (whose Internet activity is sometimes restricted as a condition of probation) but for all convicted criminals for the duration of their probation.* Someone found guilty of insider trading could be temporarily barred from all forms of e-commerce: no trading, online banking or buying things on the Internet. Or someone subjected to a restraining order would be restricted from visiting the social-networking profiles of the targeted person and his or her friends, or even searching for his or her name online.

Alas, many of these solutions will be circumvented in the age of cyber terrorism, as more and more criminals operate invisibly.

---

* This will still prove difficult to achieve, depending on the nature of the crime. Kevin Mitnick, who was an infamous computer hacker, was convicted, spent five years in prison and then, as part of his probation, was forbidden to use the Internet or a cell phone. He eventually fought the restriction through the legal system and won.

The Rise of Terrorist Hackers

How serious someone considers the threat of cyber terrorism likely depends on that person's view of hacking. For some, the image of a basement-dwelling teenager commandeering phone systems for a joy-ride endures, but hacking has developed considerably in the past decade, transformed from a hobby into a controversial mainstream activity. The emergence of "hacktivists" (politically or socially moti-vated hackers) and groups like the hacking collective Anonymous signals a maturation of message and method and hints at what we can expect in the coming years. Increasingly, hackers will find ways to organize themselves around common causes. They will conduct sophisticated attacks on whomever they deem a proper target and then publicize their successes widely. These groups will continue to demand attention from the governments and institutions they attack, and their threats may come to be taken more seriously than one might expect judging from today's activities, which mostly seem like stunts. The story of WikiLeaks, the secrets-publishing website we discussed ear-lier, and its sympathetic hacker allies is an illustrative example.

The arrest of WikiLeaks' cofounder Julian Assange in December 2010 sparked flurries of outrage around the world, particularly among the many activists, hackers and computer experts who believed his indictment on sexual-assault charges was politically motivated. Shortly thereafter, a series of cyber attacks crippled, among others, the websites for Amazon, which had revoked WikiLeaks' use of its servers, and MasterCard and PayPal, which had both stopped processing dona-tions for WikiLeaks.

This campaign, officially titled Operation Avenge Assange, was coordinated by Anonymous, a loosely knit collective of hackers and activists already responsible for a string of prominent DDoS attacks against the Church of Scientology and other targets. During Opera-tion Avenge Assange, the group vowed to take revenge on any organi-zation that lined up against WikiLeaks: "While we don't have much of an affiliation with WikiLeaks, we fight for the same reasons. We want transparency and we counter censorship. The attempts to silence WikiLeaks are long strides closer to a world where we cannot say what we think and are unable to express our opinions and ideas. We cannot let this happen. . . . This is why we intend to utilize our resources to raise awareness, attack those against and support those who are help-ing lead our world to freedom and democracy." The corporate websites were back online within several hours, but their disabling was very public and could have affected millions of customers. Most of those customers had no idea the websites were vulnerable in the first place. In other words, the hacktivists made their point. A string of global investigations followed, leading to the arrest of dozens of suspected participants in the Netherlands, Turkey, the United States, Spain and Switzerland, among other states.

Neither WikiLeaks nor groups like Anonymous are terrorist organi-zations, although some might claim that hackers who engage in activi-ties like stealing and publishing personal and classified information online might as well be. The information released on WikiLeaks put lives at risk and inflicted serious diplomatic damage.* And that's the point: Whatever lines existed between the harmless hackers and the dangerous ones (or between hackers and cyber terrorists, for that mat-ter) have become increasingly blurred in the post-9/11 era. Decentral-ized collectives like Anonymous demonstrate clearly that a collection of determined people who don't know each other, and without having met in person, can organize themselves and have a real impact in vir-tual space. In fact, no critical mass is necessary—an individual with technical prowess (computer-engineering skill, for example) can com-mandeer thousands of machines to do his bidding. What will happen in the future when there are more of these groups? Will they all fight on the side of free speech? Recent examples suggest we should begin preparing for other possibilities.

In 2011, the world met a twenty-one-year-old Iranian software engi-neer, apparently working in Tehran, who called himself Comodo-hacker. He was unusual compared to other hacktivists, who generally combat government control over the Internet, because as he told *The New York Times* via e-mail, he believed his country "should have con-trol over Google, Skype, Yahoo!, etc." He made it clear that he was intentionally working to thwart antigovernment dissidents within Iran. "I'm breaking all encryption algorithms," he said, "and giving power to my country to control all of them."

---

* At a minimum, platforms like WikiLeaks and hacker collectives that traffic in stolen clas-sified material from governments enable or encourage espionage.

Boasting aside, Comodohacker was able to forge more than five hundred Internet security certificates, which allowed him to thwart "trusted website" verification and elicit confidential or personal information from unwitting targets. It was estimated that his efforts compromised the communications of as many as three hundred thousand unsuspecting Iranians over the course of the summer. He targeted companies whose products were known to be used by dissident Iranians (Google and Skype), or those with special symbolic significance. He said he attacked a Dutch company, DigiNotar, because Dutch peacekeepers failed to protect Bosnian Muslims in Srebrenica in 1995.

Just months after Comodohacker's high-profile campaign, another ideological hacktivist from the Middle East emerged. He called himself OxOmar, claimed to live in Riyadh, Saudi Arabia, and declared that he was "one of the strongest haters of Israel" who would "finish Israel electronically." In January 2012, he hacked into a well-known Israeli sports website and redirected visitors to a site where they could download a file that contained four hundred thousand credit-card numbers (most of these were duplicates, and the total number of compromised cardholders was closer to 20,000). He claimed to represent a group of Wahhabi hackers, Group-XP, who wrote in a statement, "It will be so fun to see 400,000 Israelis stand in line outside banks and offices of credit card companies . . . [and] see that Israeli cards are not accepted around the world, like Nigerian cards." Weeks later, when the websites of Israel's El Al Airlines and its stock exchange were brought down with DoS attacks, OxOmar told a reporter that he had teamed up with a pro-Palestinian hacker group called Nightmare and that the attacks would be reduced if Israel apologized for its "genocide" against Palestinians. Israel's deputy minister of foreign affairs, Danny Ayalon, said he considered it a "badge of honor that I have been personally targeted by cyber-terrorists." He later confirmed the attacks on his Facebook page but added that hackers "will not silence us on the Internet or in any forum." Was Comodohacker really a young Iranian engineer? Did OxOmar really coordinate with another group to launch his attacks? Were these hackers individuals, or actually groups? Could either or both of these figures just be constructs of states looking to project their digital power? Any number of scenarios could be true, and therein lies the challenge of cyber terrorism in the future. Because it is very difficult to confirm the origins of cyber attacks, the target's ability to respond appropriately is compromised, regardless of who claims responsibility. This obfuscation adds a whole new dimension to misinformation campaigns, and no doubt states and individuals alike will take advantage of it. In the future, it will be harder to know who or what we are dealing with.

Sudden access to technology does not in and of itself enable radicalized individuals to become cyber terrorists. There is a technical skills barrier that, to date, has forestalled an explosion of terrorist-hackers. But we anticipate that this barrier will become less significant as the spread of connectivity and low-cost devices reaches remote places like the Afghanistan-Pakistan border region, the African Sahel and Latin America's tri-border area (Paraguay, Argentina and Brazil). Hackers in developed countries are typically self-taught, and because we can assume that the distribution of young people with technical aptitude is equivalent everywhere, this means that with time and connectivity, potential hackers will acquire the necessary information to hone their skills. One outcome will be an emergent class of virtual soldiers ripe for recruitment.

Whereas today we hear of middle-class Muslims living in Europe going to Afghanistan for terror-camp training, we may see the reverse in the future. Afghans and Pakistanis will go to Europe to learn how to be cyber terrorists. Unlike training camps with rifle ranges, monkey bars and obstacle courses, engineering boot camps could be as nondescript as a few rooms with some laptops, run by a set of technically skilled and disaffected graduate students in London or Paris. Terrorist training camps today can often be identified by satellite; cyber boot camps would be indistinguishable from Internet cafés.

Terrorist groups and governments alike will try to recruit engineers and hackers to fight for their side. Recognizing how a cadre of technically skilled strategists enhances their destructive capacity, they will increasingly target engineers, students, programmers and computer scientists at universities and companies, building out the next generation of cyber jihadists. It is hard to persuade someone to become a terrorist, given the physical and legal consequences, so surely ideology, money and blackmail will continue to play a large role in the recruitment process. Unlike governments, terrorist groups can play the antiestablishment card, which may strengthen their case among some young and disaffected hacker types. Of course, the decision to become a cyber terrorist is almost always less consequential to one's personal health than signing up for suicide martyrdom.

Culture will play an important role in where these pockets of cyber terrorism develop in the world. Deeply religious populations with distinct radicalized elements have traditionally been the most fertile ground for terrorist recruitment, and that will hold true for cyber-terrorist recruitment as well, especially as the largely disconnected parts of the world come online. To a large extent, the web experience of users is highly determined by their existing networks and immediate environment. We do not expect radical social change simply from the advent of connectivity. What we'll see instead are more communication channels, more participation and more rogue identities developing online.

And, of course, there are state sponsors of terrorism who will seek to conduct untraceable attacks. Today, Iran is one of the world's most notorious sponsors of terror groups, funneling weapons, money and supplies to groups like Hezbollah, Hamas, Palestinian Islamic Jihad, the al-Aqsa Martyrs Brigades and various militant groups in Iraq. But as cyber-terrorist efforts begin to look more fruitful, Iran will work to develop the virtual capacity of its proxies in equal measure. This means sending computer and network equipment, security packages and relevant software, but it also could mean in-person training. Iran's technical universities may well begin hosting Lebanese Shia programmers with the specific aim of integrating them into Hezbollah's emergent cyber army. Perhaps they will send them the most expensive encryption tools and hardware. Or Iran could fund technical madrassas in Hezbollah strongholds like Dahieh, Baalbek and the south of Lebanon, creating incubators where promising engineers are trained to launch cyber attacks against Israel. Perhaps instead of giving cash to Shia businessmen in Brazil to start businesses (a known tactic of the Iranian government), the regime will provide them with tablets and mobile devices carrying specialized software.

But any regime or terrorist group that recruits these hackers will assume a certain risk. While not all recruits will be young, a decent percentage will be, and not just for demographic reasons: Social scientists have long believed that certain developmental factors make young people uniquely susceptible to radicalization. (There is considerable discussion about what, precisely, those factors are, however; some believe it has to do with brain chemistry, while others argue that sociological elements in society are the driving cause.) So not only will recruiters be faced with organizing hackers, who thus far have shown a distinct resistance to formal organization, but they'll also have to deal with teenagers. As we'll discuss below, participation in a virtual-terrorism network will require inordinate discipline, not the trait most frequently associated with teenagers. Most young people are attracted to and tempted by the same things—attention, adventure, affirmation, belonging and status. Yet one mistake, or one casual boast online from a teenager hacker (or someone he knows), could unravel his entire terrorist network.

Just as counterterrorism operations today depend on intelligence sharing and military cooperation—such as that between the United States and its allies in South Asia—in the future, that bilateral support will necessarily include a virtual component. Given that many of the most radicalized countries will be among the latest arrivals to the Internet, they will need foreign support to learn how to track down terrorists online and how to use the tools newly available to them. Today, large contractors make a fortune benefiting from foreign military assistance; as bilateral efforts increasingly come to include cybersecurity elements, a range of new and established computer-security firms will benefit accordingly.

Military policies too will change in response to the threat cyber terrorists pose. Today, most of the terrorists the military chases down are in failed states or ungoverned regions. In the future, these physical safe havens will also be connected, allowing terrorists to engage in nefarious virtual acts without any fear of law enforcement. When intelligence reveals known cyber terrorists planning something dangerous, extreme measures like drone strikes will come under consideration.

Western governments will try to attract skilled hackers to their side as well. In fact, hackers and government agencies in the United States work together already, at least in matters of cybersecurity. For years, agencies like the Pentagon's Defense Advanced Research Projects Agency (DARPA) and the National Security Agency (NSA) have recruited talented individuals at venues like the computer-security conference series Black Hat and the hacker convention Def Con. In 2011, DARPA announced a new program called Cyber Fast Track (CFT), created by a former hacker turned DARPA project manager, which aimed to accelerate and streamline the cooperation between these communities. Through CFT, DARPA began awarding short-term contracts to individuals and small companies to focus on targeted network-security projects. This initiative was distinguished by its focus on smaller players and lone actors rather than big companies, and its ability to greenlight grants quickly. DARPA approved eight contracts in the first two months of the program—in other words, at lightning speed compared with the normal pace of government contracting. This process allowed groups with considerable skill who would otherwise not work with or for the government to contribute to the important work of improving cybersecurity, easily and in a time frame that reflects the immediate nature of the work. CFT was part of a shift in the agency toward "democratized, crowd-sourced innovation" championed by Regina Dugan.

We asked Dugan about the motivation behind this unconventional approach to problem-solving—after all, inviting hackers into the tent to handle sensitive security matters raises more than a few eyebrows. "There is a sense among many that hackers and Anonymous are just evildoers," she said. "What we recognized and tried to get others to embrace was that 'hacker' is a description of a talent set. Those who were declared (self-declared or otherwise) 'hackers' had something rather significant to contribute to the issue of cybersecurity, with respect to how they approach problems, the timelines on which they approach problems and their ability to execute and challenge." The success of Cyber Fast Track, she added, was a signal of the viability of that model. "I don't think that should be the only model we use," she said, "but it should absolutely be part of our approach."

More outreach to hackers and other independent computer experts should be a priority in the coming years, and we expect that Western governments will continue to try to include them, either overtly, through programs like CFT, or covertly, through the channels of intelligence agencies. Governments will push hard to acquire virtual counterparts in foreign countries to complement their undercover operators and assets active in the physical world, recruiting hackers and other technically savvy individuals to become sources and dealing with them remotely over secure online channels. There are implicit challenges associated with virtual assets, however. Despite their usefulness, there would be an absence of in-person interactions, which intelligence operatives have relied on for centuries to determine the credibility of a source. A video chat is hardly the same thing, so agencies will have to figure out how they can vet new participants effectively. Trusting virtual assets may in fact be harder than turning them.

The Terrorists' Achilles' Heel

Terrorists in the future will find that technology is necessary but high-risk. The death of Osama bin Laden, in 2011, effectively ended the era of the cave-dwelling terrorist isolated from the modern technological ecosystem. For at least five years, bin Laden hid in his mansion in Abbottabad, Pakistan, without access to the Internet or mobile phones. He had to stay off-line to stay safe. This drastically reduced his reach and influence through an al-Qaeda network that relied, at least in part, on connectivity to operate. Ironically, it was his lack of Internet access in a large urban home that helped identify him, once his courier pointed intelligence operatives in the right direction.

And while bin Laden may have evaded capture by staying off-line, even he used flash drives, hard drives and DVDs to stay informed. These tools enabled him to keep track of al-Qaeda's operations internationally and provided an efficient way for his couriers to move large amounts of data between him and various terror cells elsewhere. As long as he was at large, the information on these devices was secure, impossible to access. But when Navy SEAL Team Six raided his home, they seized his devices, getting not just the world's most wanted man but also critical information about everyone he had been in contact with.

The more likely terrorist scenario continuing into the new digital age will resemble the Mumbai attacks in 2008, when ten masked men held the city hostage in a three-day siege in which 174 people were killed and more than 300 wounded. The gunmen relied on basic consumer technologies—BlackBerrys, Google Earth and VoIP—to coordinate and conduct the attacks, communicating at a command center in Pakistan with leaders who watched live coverage of the events on satellite television and monitored the news to provide real-time tactical direction. Technology made these attacks much more deadly than they could have been otherwise, but once the last (and only surviving) gunman was captured, the information he and, critically, the leftover devices of his comrades, provided allowed investigators to follow an electronic trail to significant people and places in Pakistan that might not have otherwise been known for months, if ever.

The silver lining of cyber terrorism is that, in almost every way, its practitioners will have less room for error. Most of us have no reason to think about how differently we might interact with technology if our freedom or lives depended on erasing the tracks we leave when we go online. Cyber terrorists possess an unusually high technical awareness, but what about their friends? What about the relatives they correspond with? It is unrealistic to expect perfectly disciplined behavior from every terrorist online. Consider the nonterrorist example of John McAfee, the millionaire antivirus software pioneer who became an international fugitive after fleeing from the authorities who wanted to question him in connection with the murder of his neighbor in his adopted home country, Belize. After inviting journalists from *Vice,* an online magazine, to interview him at his secret hideout, McAfee posed for a picture with *Vice*'s editor in chief, taken with an iPhone 4S. What he—and his *Vice* interviewer—didn't know was that publishing that photo also gave away McAfee's location, since many smart phones (including the iPhone 4S) embed metadata about GPS coordinates into camera shots. All it took was one Twitter user to notice the metadata and suddenly the authorities, and the world, knew McAfee was in Guatemala, near a swimming pool at the Ranchon Mary restaurant. *Vice* should have known better (we've known about location metadata for years), but as smart phones become ever more complicated, the number of small details to keep track of compounds.

As social, professional and personal lives move increasingly to cyberspace, the interconnectedness of all digital activity increases dramatically. Computers are very good at recognizing patterns and solving needle-in-the-haystack problems, so with more data, computer algorithms can compute more precise predictions and correlations—faster and with more accuracy than any human could. Imagine a Moroccan extremist in France who has done everything possible to anonymize his smart phone from its mobile network. He has turned off geo-location, opted out of all data sharing and removes his SIM card periodically in case anyone tries to track it. He has even adopted the habit of taking the battery out of his phone as a final safeguard, knowing that when a phone is turned off, a battery retains the power to send and receive signals. His phone number is simply one of thousands, impossible to pinpoint or link to him or his location. Yet law enforcement knows he has a fondness for betting on horse races, and they know there are four off-track betting locations in his town. Using that data, they can narrow down the potential pool of numbers from thousands to the hundred or so that frequent those places. And let's say a few of his known acquaintances are not as careful with their data tracks as he is; law enforcement can then cross-reference that off-track-betting pool with the various locations of his friends. That could be all they need to do to identify his number. This type of big-data investigation was once unthinkable, but it's easy today—yet another example of humans and computers splitting duties according to their strengths. Off-line or online, our activities (and those of our friends, families and our demographic) provide intelligent computer systems with more than enough information to identify us.

It takes only one mistake or weak link to compromise an entire network. A Navy SEAL Team Six member we talked with described a top al-Qaeda commander who was exceptionally cautious around technology, always swapping phones and rarely speaking for very long. But while he was careful with his professional life, he was careless with his social one. At one point, he called a cousin in Afghanistan to say he planned to attend a wedding. That one misstep gave authorities enough information to find and capture him. Unless a terrorist is acting completely alone (which is rare), and with perfect online discipline (even rarer), there is a very good chance that somewhere in the chain of events leading up to a planned attack, he will compromise himself in some way. There are simply too many ways to reveal oneself, or be revealed, and this is very encouraging in contemplating the future of counterterrorism.

Of course, amid all of the smart and savvy cyber terrorists there will be dumb ones, too. In the trial-and-error period of connectivity growth, there will be plenty of demonstrations of inexperience that might seem laughable to those of us who grew up with the Internet.

Three years after the Canadian journalist Amanda Lindhout was kidnapped in Somalia—she was held for fifteen months by al-Shabaab extremists and finally released for a hefty ransom—her former captors contacted her on Facebook, issuing threats and inquiring about more money. Some were dummy accounts, set up with the sole purpose of harassing her further, but others appeared to be genuine personal Facebook accounts. It seems unlikely that the terrorists understood the degree to which they'd exposed themselves—not just their names and profiles, but everyone they were connected to, everything they'd written on their own and others' Facebook pages, what websites they'd "liked," and so on. Each such exposure, of course, will represent a teachable moment for other extremists, enabling them to avoid the same errors in the future.

It is estimated that more than 90 percent of people worldwide who have mobile phones keep them within three feet of themselves twenty-four hours a day. There is no reason to believe this won't be true for extremists. They might adopt new routines that help protect them—like periodically removing the battery from their phones—but they won't stop using them altogether. This means that counterterrorist raids by militaries and law enforcement will result in better outcomes: capture the terrorist, capture his network. Interrogations post-capture will remain important, but each device used by a terrorist—mobile phones, storage drives, laptops and cameras—will be a potential gold mine. Commandeering a captured terrorist's devices with the rest of his network unaware will lead his cohorts to unwittingly disclose sensitive information or locations. Additionally, the devices might contain content that can be used to expose hypocrisy in a terrorist's public persona, as American officials did when they revealed that the computer files taken from Osama bin Laden's compound contained a large stash of pornographic videos. Of course, once this vulnerability becomes apparent, more sophisticated terrorists will combat it by having an abundance of technology with misleading information on it. Deliberately storing personal details about rivals or enemies on devices that find their way into the hands of law enforcement will be a useful form of sabotage.

No Hidden People Allowed

As terrorists develop new methods, counterterrorism strategists will adapt accordingly. Imprisonment may not be enough to contain a terror network. Governments may determine, for example, that it is too risky to have citizens "off the grid," detached from the technological ecosystem. To be sure, in the future, as now, there will be people who resist adopting and using technology, people who want nothing to do with virtual profiles, online data systems or smart phones. Yet a government might suspect that people who opt out completely have something to hide and thus are more likely to break laws, and as a counterterrorism measure, that government will build the kind of "hidden people" registry we described earlier. If you don't have any registered social-networking profiles or mobile subscriptions, and online references to you are unusually hard to find, you might be considered a candidate for such a registry. You might also be subjected to a strict set of new regulations that includes rigorous airport screening or even travel restrictions.

In a post-9/11 world, we can already see signs that even countries with strong civil-liberties foundations jettison citizen protections in favor of a system that enhances homeland surveillance and security. That will only accelerate. After some cyber-terrorist successes, it will be easier to persuade people that the sacrifices involved—essentially, a heightened level of governmental monitoring of online activity—are worth the peace of mind they will bring. The collateral damage in this scenario, besides the persecution of a small number of harmless hermits, of course, is the danger of the occasional abuse or poor judgment by government stewards. This is yet another reason it will be so important to fight for privacy and security in the future.

The push-pull between privacy and security in the digital age will become even more prominent in the coming years. The authorities responsible for locating, monitoring and capturing dangerous individuals will require massive, highly sophisticated data-management systems to do so. Despite everything individuals, corporations and dedicated nonprofit groups are doing to protect privacy, these systems will inevitably include volumes of data about non-terrorist citizens—the questions are how much and where. Currently, most of the information that governments collect on people—their addresses,

ID numbers, police records, mobile-phone data—is siloed in separate places (or is not even digitized yet in some countries). Its separation ensures a degree of privacy for citizens but creates large-scale inefficiencies for investigators.

This is the "big data" challenge that government bodies and other institutions around the world are facing: How can intelligence agencies, military divisions and law enforcement integrate all of their digital databases into a centralized structure so that the right dots can be connected without violating citizens' privacy? In the United States, for example, the FBI, State Department, CIA and other government agencies all use different systems. We know computers can find patterns, anomalies and other relevant signifiers much more efficiently than human analysts can, yet bringing together disparate information systems (passport information, fingerprint scans, bank withdrawals, wiretaps, travel records) and building algorithms that can efficiently cross-reference them, eliminate redundancy and recognize red flags in the data is an incredibly difficult and time-consuming task.

Difficult does not mean impossible, however, and all signs point toward these comprehensive integrated information systems' becoming the standard for modern, wealthy states in the near future. We had the opportunity to tour the command center for Plataforma México, Mexico's impressive national crime database and perhaps the best model of an integrated data system operating today. Housed in an underground bunker in the Secretariat of Public Security compound in Mexico City, this large database integrates intelligence, crime reports and real-time data from surveillance cameras and other inputs from agencies and states across the country. Specialized algorithms can extract patterns, project social graphs and monitor restive areas for violence and crime as well as for natural disasters and other civilian emergencies. The level of surveillance and technological sophistication of Plataforma México that we saw is extraordinary—but then, so are the security challenges that Mexican authorities face. Therein lies the challenge looking ahead: Mexico is the ideal location for a pilot project like this because of its entrenched security problems, but once the model has been proven, what is to stop other states with less justifiable motivations from building something similar? Surely other governments can play the security card and insist that such a sophisticated platform is necessary; what might stop them?

In the early 2000s, following the September 11 terrorist attacks, something similar was proposed in the United States. The Defense Department set up the Information Awareness Office and green-lit the development of a program called Total Information Awareness (TIA). Pitched as the ultimate security apparatus to detect terrorist activity, TIA was designed and funded to aggregate all "transactional" data— including bank records, credit-card purchases and medical records— along with other bits of personal information to create a centralized and searchable index for law enforcement and counterterrorist agencies. Sophisticated data-mining technologies would be built to detect patterns and associations, and the "signatures" that dangerous people left behind would reveal them in time to prevent another attack.

As details of the TIA program leaked out to the public, a range of vocal critics emerged from both the right and the left, warning about the potential costs to civil liberties, privacy and long-term security. They zeroed in on the possibilities of abuse of such a massive information system, branding the program "Orwellian" in scope. Eventually, a congressional campaign to shut TIA down resulted in a provision to deny all funds for the program in the Senate's 2004 defense appropriations bill. The Information Awareness Office was shuttered permanently, though some of its projects later found shelter in other intelligence agencies in the government's sprawling homeland-security sector.

Fighting for privacy is going to be a long, important struggle. We may have won some early battles, but the war is far from over. Generally, the logic of security will always trump privacy concerns. Political hawks merely need to wait for some serious public incident to find the political will and support to push their demands through, steamrolling over the considerations voiced by the doves, after which the lack of privacy becomes normal. With integrated information platforms like these, adequate safeguards for citizens and civil liberties must be firmly in place from the onset, because once a serious security threat appears, it is far too easy to overstep. (The information is already there for the taking.) Governments operating surveillance platforms will surely violate restrictions placed on them (by legislation or legal ruling) eventually, but in democratic states with properly functioning legal systems and active civil societies, those errors will be corrected, whether that means penalties for perpetrators or new safeguards put into place.

Serious questions remain for responsible states. The potential for misuse of this power is terrifyingly high, to say nothing of the dangers introduced by human error, data-driven false positives and simple curiosity. Perhaps a fully integrated information system, with all manner of data inputs, software that can interpret and predict behavior and humans at the controls is simply too powerful for anyone to handle responsibly. Moreover, once built, such a system will never be dismantled. Even if a dire security situation were to improve, what government would willingly give up such a powerful law-enforcement tool? And the *next* government in charge might not exhibit the same caution or responsibility with its information as the preceding one. Such totally integrated information systems are in their infancy now, and to be sure they are hampered by various challenges (like consistent data-gathering) that impose limits on their effectiveness. But these platforms will improve, and there is an air of inevitability around their proliferation in the future. The only remedies for potential digital tyranny are to strengthen legal institutions and to encourage civil society to remain active and wise to potential abuses of this power.

A final note on digital content as we discuss its uses in the future: As online data proliferates and everyone becomes capable of producing, uploading and broadcasting an endless amount of unique content, verification will be the real challenge. In the past few years, major news broadcasters have shifted from using only professional video footage to including user-generated content, like videos posted to YouTube. These broadcasters typically add a disclaimer that the video cannot be independently verified, but the act of airing it is, in essence, an implicit verification of its content. Dissenting voices may claim that the video has been doctored, or is somehow misleading, but those claims when registered get a fraction of the attention and are often ignored. The trend toward trusting unverified content will eventually spur a movement toward more rigorous, technically sound verification.

Verification, in fact, will become more important in every aspect of life. We explored earlier how the need for verification will come to shape our online experiences, requiring better protections against identity theft, with biometric data changing the security landscape. Verification will also play an important role in determining which terrorist threats are actually valid. To avoid identification, most extremists will use multiple SIM cards, multiple online identities and a range of obfuscating tools to cover their tracks. The challenge for law enforcement will be finding ways to handle this information deluge without wasting man-hours on red herrings. Having "hidden people" registries in place will reduce this problem for authorities but will not solve it.

Because the general public will come to prefer, trust, depend on or insist on verified identities online, terrorists will make sure to use their own verified channels when making claims. And there will be many more ways to verify the videos, photos and phone calls that extremist groups use to communicate. Sharing a photograph of hostages with fresh daily newspapers will become an antiquated practice—the photo itself is the proof of when it was taken. Through digital forensic techniques like checking the digital watermarks, IT experts can verify not only when, but where and how.

This emphasis on verified content, however, will require terrorists to make good on their threats. If a known terrorist does not do so, the subsequent loss of credibility will hurt his and his group's reputation. If al-Qaeda were to release an audio recording proving that one of its commanders survived a drone attack, but forensic computer experts using voice-recognition software determined that someone else's voice was on the tape, it would weaken al-Qaeda's position and embolden its critics. Each verification challenge would chip away at the grandiose image that many extremist groups rely on to raise funds, recruit and instill fear in others. Verification can therefore be a tremendous tool in the fight against violent extremism.

The Battle for Hearts and Minds Comes Online

While it's true that effective hackers and computer experts will enhance terror groups' capabilities, the broad foundation of recruits will, like today, be basic foot soldiers. They'll be young and undereducated, and they'll have grievances that extremists exploit to their own advantage. We believe that the most pivotal shift in counterterrorism strategy in the future will not concern raids or mobile monitoring, but instead will focus on chipping away at the vulnerability of these at-risk populations through technological engagement.

An estimated 52 percent of the world's population is under the age of thirty, and the vast majority are what we could call "socioeconomically at risk," living in urban slums or poorly integrated immigrant communities, in places with unreliable rule of law and limited economic opportunity. Poverty, alienation, humiliation, lack of opportunity and mobility, and just simple boredom make these young populations highly susceptible to the influence of others. Set against a backdrop of repression and in a subculture that promotes extremism, their grievances foster their radicalization. This is as true for the undereducated slum kid as it is for university students who see no opportunities awaiting them on the other side of their degree.

At Google Ideas, we've studied radicalization around the world, particularly with an eye toward the role that communication technologies can play.* It turns out that the radicalization process for terrorists is not very different from what we see with inner-city gangs or other violent groups, like white supremacists. At our Summit Against Violent Extremism in June 2011, we brought together more than eighty former extremists to discuss why individuals join violent organizations, and why they leave them. Through open dialogue with the participants, who, between them, represented religious extremists, violent nationalists, urban gangs, far-right fascists and jihadist organizations, we learned that similar motivations exist across all these groups, and that religion and ideology play less of a role than most people think. The reasons people join extremist groups are complex, often having to do more with the absence of a support network, the desire to belong to a group, to rebel, to seek protection or to chase danger and adventure.

There are far too many young people who share these sentiments. What's new is that large numbers of them will air their grievances online in ways that advertently or inadvertently advertise themselves to terrorist recruiters. What radicalized youth seek through virtual connections grows out of their experience in the physical world—abandonment, rejection, isolation, loneliness and abuse. We can figure out a great deal about them in the virtual world, but in the end, real

* Google, like many other companies, builds free tools that anyone can use. Because of this, the company is continuously working to understand how to mitigate the risks that hostile individuals and entities will use these tools to cause harm.

de-radicalization requires group meetings and a lot of support, therapy and meaningful alternatives in the physical world.

Words and speeches against violent extremism will not be enough in the battle for young hearts and minds. Military force will not do the job, either. Governments have been largely successful at capturing and killing existing terrorists but less effective at stemming the flow of recruits. As General Stanley McChrystal, the former U.S. and NATO commander in Afghanistan, told *Der Spiegel* in 2010, "What defeats terrorism is really two things. It's rule of law and then it's opportunity for people. So if you have governance that allows you to have rule of law, you have an environment in which it is difficult to pursue terrorism. And if you have an opportunity for people in life, which includes education and the chance to have a job, then you take away the biggest cause of terrorism. So really, the way to defeat terrorism is not military strikes, it's going after the basic conditions."

McChrystal's insight spotlights an opportunity for technology enthusiasts and companies alike, because what better way is there to improve a population's quality of life than boosting its connectivity? The gains that communication technologies produce for communities—economic opportunity, entertainment, freedom of information, greater transparency and accountability—all contribute to the antiradicalization mission. Once a large segment of a population is online, it will be possible to mobilize the local virtual community to reject terrorism and demand accountability and action from its leaders. There will be more voices speaking out against extremism than for it, and while technology may extend the reach of fanatics, it will be impossible to preach one way of thinking without encountering some interference. All of the things that come with an active virtual sphere— more discussion, more points of view, more counternarratives—can introduce doubt and promote independent thinking among these young, malleable populations. Of course, all of this would fall on more receptive ears if connectivity led to job creation.

The most potent antiradicalization strategy will focus on the new virtual space, providing young people with content-rich alternatives and distractions that keep them from pursuing extremism as a last resort. This effort should be large and involve stakeholders from every background—the public sector, private companies, partnerships between local actors and activists abroad. Mobile technology, particularly, will play the dominant role in this campaign, since the majority of people coming online will do so through their handsets. Phones are personalized and powerful platforms, status symbols that their users rely on and value deeply. Reaching disaffected youth through their mobile phones is the best possible goal we can have.

Western companies and governments will not be the ones to develop the bulk of the new content. The best solutions will be hyperlocal, designed and supported by people with intimate knowledge of the immediate environment. Building platforms that we merely *hope* alienated youth will like and use is the equivalent of dropping propaganda flyers from an airplane.

Outsiders don't have to develop the content; they just need to create the space. Wire up the city, give people basic tools and they'll do most of the work themselves. A number of technology companies have developed start-up kits for people to build applications on top of their platforms; Amazon Web Services and Google App Engine are two examples, and there will be many others. Creating space for others to build the businesses, games, platforms and organizations they envision is a brilliant corporate maneuver, because it ensures that a company's products are used (boosting brand loyalty, too) while the users actually build and operate what *they* want. Somalis will build apps that are effective antiradicalization tools to reach other Somalis; Pakistanis will do the same for other Pakistanis. There will be more opportunities for local people to build small businesses and create outlets for youth at the same time. The key is to simply enable people to adapt products in ways that fit their needs and don't require complex technological expertise.

Public-private partnerships with local activists and people of influence will drive this process. Companies should also look to partnering with local groups to develop content. Ideally, what emerges is a range of content, platforms and applications that speak to each community distinctly yet share technological or structural components enabling them to be mimicked in other places. If the causes of radicalization are similar everywhere, the remedies can be, too.

Technology companies are uniquely positioned to lead this effort internationally. Many of the most prominent ones have all the values of a democratic society with none of the baggage of being a government—they can go where governments can't, speak to people off the diplomatic radar and operate in the neutral, universal language of technology. Moreover, this is the industry that produces video games, social networks and mobile phones—it has perhaps the best understanding of how to distract young people of any sector, and kids are the very demographic being recruited by terror groups. The companies may not understand the nuances of radicalization or the differences between specific populations in key theaters like Yemen, Iraq and Somalia, but they do understand young people and the toys they like to play with. Only when we have their attention can we hope to win their hearts and minds.

Moreover, due to technology companies' involvement in security threats—their products are being used by terrorists—the public will eventually demand that they do more in the fight against extremism. This means not only improving their products and protecting users with strict policies regarding content and security, but also taking a public stand. Just as the capitulation of MasterCard and PayPal to political pressure in the WikiLeaks saga convinced many activists that the companies took sides, inaction on the part of technology companies will be considered indefensible to some. Fairly or otherwise, companies will be held responsible for destructive uses of their products. Companies will reveal their personalities and core values according to how they rise to meet these challenges. Empty words will not pacify an informed public.

We can already see early steps in this direction, as some companies make clear statements in policy or procedure. At YouTube, there is the challenge of content volume. With more than four billion videos viewed daily (and sixty hours of video uploaded each minute), it is impossible for the company to screen that content for what is considered inappropriate material, like advocating terrorism. Instead YouTube relies on a process in which users flag content they consider inappropriate; the video in question goes to a YouTube team for review, and it is taken down if it violates company policies. Eventually, we will see the emergence of industry-wide standards. All digital platforms will forge a common policy with respect to dangerous extremist videos online, just as they have coalesced in establishing policies governing child pornography. There is a fine line between censorship and security, and we must create safeguards accordingly. The industry will work as a whole to develop software that more effectively identifies videos with terrorist content. Some in the industry may even go so far as employing speech-recognition software that registers strings of key-words, or facial-recognition software that identifies known terrorists.

Terrorism, of course, will never disappear, and it will continue to have a destructive impact. But as the terrorists of the future are forced to live in both the physical and the virtual world, their model of secrecy and discretion will suffer. There will be more digital eyes watching, more recorded interactions, and, as careful as even the most sophisticated terrorists are, even they cannot completely hide online. If they are online, they can be found. And if they can be found, so can their entire network of helpers.

In this chapter we have explored the darkest ways that individuals will seek to violently disrupt our future world, but given that conflict and war are as much a part of human history as society itself, how will states and political movements engage in these activities to achieve their aims? We'll now explore this question by imagining how conflict, combat and intervention are affected in a world where almost everyone is online.

CHAPTER 6

The Future of Conflict, Combat and Intervention

Never before have we been so aware of so many conflicts around the world. The accessibility of information about atrocities anywhere—the stories, the videos, the photos, the tweets—can often make it seem as though we live in an exceptionally violent time. But as the newspaper adage goes, "If it bleeds, it leads." What has changed is not how many conflicts there are, but how visible they've become.

If anything, we're more peaceful than we've ever been, with the amount of violence in human societies declining precipitously in the past several centuries due to developments like strong states (which monopolize violence and institute the rule of law), commerce (other people become more valuable alive than dead) and expanded international networks (which demystify and humanize the Other). As the psychologist Steven Pinker explains in *The Better Angels of Our Nature*, his excellent and comprehensive survey of this trend, historical exogenous forces like these "favor our peaceable motives" like empathy, moral sense, reason and self-control, which "orient [us] away from violence and toward cooperation and altruism." Once conscious of this shift, Pinker remarks, "The world begins to look different. The past seems less innocent; the present less sinister."

Surely "connectivity" would belong in Pinker's list of forces had he written his book fifty years hence, because the new level of visibility that perpetrators of violence face in a connected world and all that it portends will greatly weaken any incentives for violent action and alter the calculus of political will to commit crimes as well as stop them.

Nevertheless, conflict, wars, violent border skirmishes and mass atrocities will remain a part of human society for generations to come even as they change form in accordance with the technological age. Here we explore the ways in which different elements of conflict—the buildup of discrimination and persecution, combat and intervention—will change in the coming decades in response to these new possibilities and penalties.

Fewer Genocides, More Harassment

The origins of violent conflicts are far too complex to have a single root cause. But one well-understood trigger that will change substantially in the new digital age is the systematic discrimination or persecution of minority groups, during which targeted communities become the victims of grave violence or themselves become perpetrators of retaliatory acts. We believe that, in the future, massacres on a genocidal scale will be harder to conduct, but discrimination will likely worsen and become more personal. Increased connectivity within societies will provide practitioners of discrimination, whether they are official groups or ones led by citizens, with entirely new ways to marginalize minorities and other disliked communities, whose own use of technology will make them easier to target.

Governments that are used to repressing minorities in the physical world have a whole new set of options in the virtual world, and those that figure out how to combine their policies in both worlds will be that much more effective at repression. Should the government of a connected country wish to harass a particular minority community in the future, it will find a number of tactics immediately available. The most basic would be to simply erase content about that group from the country's Internet. States with strong filtering systems will find this easy, since the ISPs could just be required to block all sites containing certain keywords, and to shut down sites with prohibited content. To scrub the lingering references to the group on sites like Facebook and YouTube, the state could adopt an approach similar to China's policy of active censorship, where censors automatically shut down the connection whenever a prohibited word is sighted.

The Chinese government might well target the Uighur minority in western China. Concentrated in the restive Xinjiang region, this mostly Muslim Turkic ethnic group has long seen tensions flare with the majority Han Chinese, and separatist movements in Xinjiang have been responsible for a series of failed uprisings in the past several years. Though small, the Uighur population has caused countless headaches in Beijing, and it's no stretch of the imagination to think that the government could move from censoring Uighur-related episodes (like the 2009 Ürümqi riots) to eliminating all Uighur content online.

States might view this kind of action as a political imperative, an effort to mitigate the internal threats to stability by simply erasing them. Information about the groups would remain available outside a country's Internet space, of course, but internally it would vanish. This would be intended both to humiliate the group by negating its very existence and also to isolate it further from the rest of the population. The state could persecute the group with greater impunity, and in time, if the censorship was thorough enough, future generations of majority groups could grow up with barely any awareness of the minority group or the issues associated with it. Erasing content is a quiet maneuver, difficult to quantify and unlikely to set off alarm bells, because such efforts would have small tangible impact while remaining symbolically and psychologically disparaging to the groups most affected. And even if a government were to get "caught" somehow, and shown to be deliberately blocking minority-specific content, officials would probably justify their actions on security grounds or blame them on computer glitches or infrastructure failures.

If a government wanted to go further than content control, and escalate its discriminatory policies to full-blown persecution online, it could find ways to limit a given group's access to the Internet and its services. This might sound trivial in comparison with the physical harassment, random arrests, acts of violence, and economic and political strangulation that persecuted groups around the world experience today. But as connectivity spreads, Internet service and mobile devices offer vital outlets for individuals to transcend their current environment, connecting them with information, jobs, resources, entertainment and other people. Excluding oppressed populations from participating in the virtual world would be a very drastic and damaging policy, because in important ways they'd be left out and left behind, unable to tap into any of the opportunities for growth and prosperity that we see connectivity bringing elsewhere. As banking, salaries and payment transactions move increasingly onto online platforms, exclusion from the Internet will severely curtail people's economic prospects. It would be far more difficult to access one's money, to pay by credit card or get a loan.

Already, the Romanian government deliberately excludes some 2.2 million ethnic Roma from the same opportunities as the rest of the population, a policy manifested in separate education systems, economic exclusion in the form of hiring discrimination and unequal access to health and medical benefits (not to mention a heavy social stigma). Current statistics on the Roma's level of access to technology are hard to come by—many Roma fail to register themselves as such on government surveys for fear of persecution—but as we've made clear, connected Roma will find ways to improve their circumstances. The Roma might even consider pursuing virtual statehood of some kind in the future.

But if the Romanian government decided to extend its policies toward the Roma into the online world, nearly all of those opportunities would evaporate. Technological exclusion could take many forms, depending on how much control the state has and how much pain it wants to cause. If it required all citizens to register their devices and IP addresses (many governments already require mobile devices to be registered) or maintained a "hidden people" registry, Romanian authorities using that data would find it easy to block the Roma's access to news, outside information and platforms with economic or social value. These users would suddenly find themselves unable to reliably access their own personal data or their online banking services; they would confront error messages or seem to have egregiously slow connection speeds. Using its power over the country's telecommunications infrastructure, the government could instigate dropped calls, jam phone signals in certain neighborhoods or occasionally short-circuit the Roma's connections to the Internet. Perhaps the government, working with private-sector distributors, could engineer the sale of defective devices to Romany individuals (selling to them through compromised trusted intermediaries), distributing laptops and mobile phones riddled with bugs and back doors that would allow the state to input malicious code at a later date.

Rather than a systematic campaign to cut access (which would incur unwelcome scrutiny), the Romanian government would need only to implement these blockages randomly, frequently enough to harass the group itself but intermittently enough to allow for plausible denials. The Roma, of course, could find imperfect technological workarounds that enabled basic connectivity, but ultimately the blockages would be sufficiently disruptive that even intermittent access couldn't replace what was lost. Over a long enough period, a dynamic like this might settle into a kind of virtual apartheid, with multiple sets of limitations on connectivity for different groups within society.

Electronically isolating minority groups will become increasingly prevalent in the future because states have the will to do so, and they have access to the data that enables it. Such initiatives might even start as benign programs with public support, then over time transform into more restrictive and punitive policies as power shifts in the country. Imagine, for example, if the ultra-Orthodox contingent in Israel lobbied for the creation of a white-listed "kosher Internet," where only preapproved websites were allowed, and their bid was successful—after all, the thinking might be, creating a special Internet lane for them is not unlike forming a special "safe" list of Internet sites for children.* Years later, if the ultra-Orthodox swept the elections and took control of the government, their first decision might be to make all of Israel's Internet "kosher." From that position, they would have the opportunity to restrict access even further for minority groups within Israel.

The most worrisome result of such policies is how vulnerable these restrictions would make targeted groups, whose lifelines could literally be cut. If limited access were to be a precursor to physical harassment or state violence by compromising a group's ability to send out alert signals, it would also strip victims of their ability to document the abuse or destruction afterward. Soon it may be possible to say that what happens in a digital vacuum, in effect, doesn't happen.

In countries where governments are targeting minority or repressed groups in this way, an implicit or explicit arrangement between some citizens and states will emerge, whereby people trade information or obedience in exchange for better access. Where noticeable coopera-

---

* If such an exception was made for the Israeli ultra-Orthodox on religious grounds, what kind of precedent would it set? What if the ultraconservative Salafis in Egypt followed suit, demanding a special white-listed Internet?

tion with the government is demonstrated, the state will grant those individuals faster connections, better devices, protection from online harassment or a broader range of accessible Internet sites. An artist and father of six living in Saudi Arabia's Shiite minority community may have no desire to become an informant or sign a government pledge to stay out of political affairs, but if he calculates that that cooperation means a more reliable income for himself or better educational opportunities for his children, his resolve might well weaken. The strategy of co-opting potentially restive minority groups by playing to their incentives is as old as the modern state itself; this particular incarnation is merely suited for our digital age.

Neither of these tactics—erasing content and limiting access—is the purview of states alone. Technically capable groups and individuals can pursue virtual discrimination independently of the government. The world's first virtual genocide might be carried out not by a government but by a band of fanatics. Earlier, we discussed how extremist organizations will venture into destructive online activities as they develop or acquire technological skills, and it follows that some of those activities will echo the harassment described above. This goes for lone-wolf zealots, too. It's not hard to imagine that a rabidly anti-Muslim activist with strong technical skills might go after his local Muslim community's websites, platforms and media outlets to harass them. This is the virtual equivalent of defacing their property, breaking into their businesses and shouting at them from street corners. If the perpetrator is exceptionally skilled, he will find ways to limit the Muslims' access by targeting certain routers to shut them down, sending out jamming signals in their neighborhoods or building computer viruses that disable their connections.

In fact, virtual discrimination will suit some extremists better than their current options, as a former neo-Nazi leader and current anti-hate activist named Christian Picciolini told us. "Online intimidation by hate groups or extremists is more easily perpetrated because the web dehumanizes the interaction and provides a layer of anonymity and 'virtual' disconnection," he explained. "Having the Internet as an impersonal buffer makes it easier for the intimidator to say certain harmful things that might not normally be said face-to-face for fear of peer judgment or persecution. Racist rhetoric rightfully carries a certain social stigma against the general population, but online, words can be said without being connected to the one saying [them]." Picciolini expects virtual harassment by hate groups to increase significantly in the coming years, since "the consequences of online discrimination seem less audacious to the offender, and therefore [harassment will] happen more frequently and to a more vehement degree."

In the past, physical and legal exclusion was the dominant tactic used by the powerful in conflict-prone societies, and we believe that virtual exclusion will come to join (but not surpass) that tool kit. When the conditions become unbearable, as throughout history, the sparks of conflict will ignite.

Multidimensional Conflict

Misinformation and propaganda have always been central features of human conflict. Julius Caesar filled his famous account of the Gallic Wars (58 B.C.–50 B.C.) with titillating reports of the vicious barbarian tribes he'd fought. In the fog of competing narratives, determining the "good" and "bad" actors in a conflict is a critical yet often difficult task, and it will become even more challenging in the new digital age. In the future, marketing wars between groups will become a defining feature of conflict, because all sides will have access to electronic platforms, tools and devices that enhance their storytelling abilities for audiences at home and abroad. We saw this unfold during the November 2012 conflict between Israel and Hamas, when the terrorist organization launched a grassroots marketing war that flooded the virtual world with graphic photos of dead women and children. Hamas, which thrives on a public that is humiliated and demoralized, was able to exploit the larger number of casualties in Gaza. Israel, which focuses more on managing national morale and reducing ambiguity around its actions, countered by utilizing an @IDFSpokesperson Twitter handle, which included tweets like "Video: IDF pilots wait for area to be clear of civilians before striking target youtube.com /watch?v=G6a112wRmBs ... #Gaza." But the reality of marketing wars are that the side which is happy to glorify death and use it for propaganda will often gain wider-spread sympathy, especially as a larger and less-informed audience joins the conversation. Hamas's propaganda tactics were not new, but the growing ubiquity of platforms such as YouTube, Facebook and Twitter made it possible for them to reach a much larger and non-Arabic-speaking audience in the West, who with each tweet, like and plus-one amplified Hamas's marketing war.

Groups in conflict will try to destroy each other's digital marketing capabilities before a conflict even starts. Few conflicts are clearly black-and-white at the end—let alone when they start—and this near-equivalency in communications power will greatly affect how civilians, leaders, militaries and the media deal with conflict. What's more, the very fact that *anyone* can produce and share his or her version of events will actually nullify many claims; with so many conflicting accounts and without credible verification, all claims become devalued. In war, data management (compiling, indexing, ranking and verifying the content emanating from a conflict zone) will shortly succeed access to technology as the predominant challenge.

Modern communication technologies enable both the victims and the aggressors in a given conflict to cast doubt on the narrative of the other side more persuasively than with any media in history. For states, the quality of their marketing might be all that lies between staying in power and facing a foreign intervention. For civilians trapped in a town under siege by government forces, powerful amateur videos and real-time satellite mapping can counter the claims of the state and strongly suggest, if not prove, that it is lying. Yet in a situation like the 2011 violence in Côte d'Ivoire (where two sides became locked in a violent battle over contested electoral results), if both parties have equally good digital marketing, it becomes much harder to discern what is really happening. And if neither side is fully in control of its marketing (that is, if impassioned individuals outside the central command produce their own content), the level of obfuscation rises even more.

For outsiders looking in, already difficult questions, like who to speak with to understand a conflict, who to support in a conflict and how best to support them, become considerably more complicated in an age of marketing wars. (This is particularly true when not many outsiders speak the local language, or in the absence of standing alliances, like between NATO countries or the SADC countries, the Southern African Development Community.) Critical information needed to make those decisions will be buried beneath volumes of biased and conflicting content emanating from the conflict zone. States rarely intervene militarily unless it is very clear what is taking place, and even then they often hesitate for fear of the unforeseen physical consequences and the scrutiny in the twenty-four-hour news cycle.*

Marketing wars within a conflict abroad will have domestic political implications, too. If the bulk of the American public, swayed by one side's emotionally charged videos, concludes that intervention in a given conflict is a moral necessity, but the U.S. government's intelligence suggests that those videos aren't reflective of the real dynamics in the conflict, how should the administration respond? It can't release classified material to justify its position, but neither can it effectively counter the narrative embraced by the public. If both sides present equally persuasive versions, outside actors become frozen in place, unable to take a step in any direction—which might be the exact goal of one of the parties in the conflict.

In societies susceptible to ethnic or sectarian violence, marketing wars will typically begin long before there is a spark that ignites any actual fighting. Connectivity and virtual space, as we've shown, can often amplify historical and manufactured grievances, strengthening the dissonant perspectives instead of smoothing over their inaccuracies. Sectarian tensions that have lain somewhat dormant for years might reignite once people have access to an anonymous online space. We've seen how religious sensitivities can become inflamed almost instantaneously when controversial speech or images reach the Internet—the Danish cartoon controversy in 2005 and violent demonstrations over the *Innocence of Muslims* video in 2012 are just a couple of many prominent examples—and it's inevitable that online space will create more ways for people to offend one another. The viral nature of incendiary content will not allow an offensive act in any part of the world to go unnoticed.

Marketing is not the same thing as intelligence, of course. Early attempts at digital marketing by groups in conflict will be little more than crude propaganda and misinformation transferred to a virtual

---

* In policy circles, this is known as the CNN Effect, and is most frequently associated with the 1992–1993 U.S. intervention in Somalia. It's widely believed that the images broadcast on television of starving and desperate Somalis prompted George H. W. Bush to send in military forces, but when, on October 3, 1993, eighteen Army Rangers and two Malaysian coalition partners were killed and the images of one of the Americans dragged through the streets in Mogadishu reached the airwaves, the American forces were withdrawn.

platform. But over time, as these behaviors are adopted around the world by states and individuals, the aesthetic distance between intelligence and marketing will close. States will have to be careful not to mistake one for the other. Once groups are wise to what they need to produce in order to generate a specific response, they will be able to tailor their content and messaging accordingly.

Those with state resources will have the upper hand in any marketing war, but never the exclusive advantage. Even if the state controls many of the means of production—the cell towers, the state media, the ISPs—it will be impossible for any party to have a complete information monopoly. When all it takes to shoot, edit, upload and disseminate user-generated content is a palm-sized phone, a regime can't totally dominate. One video captured by a shaky mobile-phone camera during the postelection protests in Iran in 2009 galvanized the opposition movement: the famous "Neda video." Neda Agha-Soltan was a young woman living in Tehran who while parked on a quiet side of the street at an antigovernment protest stepped out of her car to escape the heat and was shot in the heart by a government sniper from a nearby rooftop. Amazingly, the entire incident was caught on someone's mobile phone. While members of the crowd attempted to revive Neda, others began filming her on their phones as well. The videos were passed between Iranians, mostly through the peer-to-peer platform Bluetooth, since the regime had blocked mobile communications in anticipation of the protests; they found their way online and went viral. Around the world, observers were galvanized to speak out against the Iranian regime while protesters in Iran marched, calling for justice for Neda. All of this significantly ratcheted up the global attention paid to a protest movement the regime was desperately trying to stop.

Even in the most restrictive societies, places where spyware and virtual harassment and pre-compromised mobile phones are rampant, some determined individuals will find a way to get their messages out. It might involve smuggling SIM cards, rigging mesh networks (essentially, a wireless collective in which all devices act like routers, creating multiple nodes for data traffic instead of a central hub) or distributing "invisible" phones that are designed to record no communications (perhaps by allowing all calls to be voice over IP) and that allow anonymous use of Internet services. All state efforts to curtail the spread of an in-demand technology fail; it's merely a question of when. (This is true even for the persecuted minorities whose government tries to exclude them from the Internet.) Long before the Neda video, Iran tried to ban satellite-television dishes; their mandate was met with an *increase* in satellite adoption among the Iranian public. Today, the illegal satellite market in Iran is among the largest per capita in the world and even some members of the regime profit from black market sales.

The 1994 Rwandan genocide, a high-profile conflict from the pre-digital age that claimed the lives of 800,000 people, demonstrates what a difference proportionate marketing power makes. In 1994, while Hutus, Tutsis and Twa all owned radios, only Hutus owned radio stations. With no means of amplifying their voices, Tutsis were powerless against the barrage of propaganda and hate speech building on the airwaves. When Tutsis tried to operate their own radio station, the Hutu-dominated government identified these operators, raided their offices and made arrests. If the minority Tutsi population in the years leading up to the 1994 genocide had had the powerful mobile devices we have today, perhaps a narrative of doubt could have been injected into Rwandan public discourse, so that some ordinary Hutu civilians would not have found the anti-Tutsi propaganda sufficiently compelling to lead them to take up arms against their fellow Rwandans. The Tutsis would have been able to broadcast their own content from handsets, while on the move, without having to rely on government approval or intermediaries to develop and disseminate content. During the genocide, the Hutu radio stations announced names and addresses of people who were hiding—one can only imagine what a difference an alternative communications channel, like encrypted peer-to-peer messaging, might have made.

Despite potential gains, there will be longer-term consequences to this new level playing field, though we cannot predict what will be lost when traditional barriers are removed. Misinformation, as mentioned above, will distract and distort, leading all actors to misinterpret events or miscalculate their response. Not every brutal crime committed is part of a systematic slaughter of an ethnic or religious group, yet it can be incorrectly cast as such with minimal effort. Even in domestic settings, misinformation can present a major problem: How should a local government handle an angry mob at city hall demanding justice for a manipulated video? Governments and authorities will face questions like these repeatedly, and only some of the answers they give will be pacifying.

The best and perhaps only reply to these challenges is digital verification. Proving that a photo has been doctored (by checking the digital watermark), or that a video has been selectively edited (by crowd-sourcing the whole clip to prove parts are missing), or that a person shown to be dead is in fact alive (by tracking his online identity) will bring some veracity in a hyper-connected conflict. In the future, a witness to a militia attack in South Sudan will be able to add things like digital watermarks, biometric data and satellite positioning coordinates to add heft to his claims, useful for when he shares the content with police or the media. Digital verification is the next obvious stage of the process. It already occurs when journalists and government officials cross-check their sources with other forms of information. It will be even easier and more reliable when computers do most of the work.

Teams of international verification monitors could be created, dispatched to conflicts where there is a significant dispute about the digital narratives emerging. Like the Red Cross, verification monitors would be seen as neutral agents, in this case highly capable ones technically.* (They need not be deployed to the actual conflict zone in every case—their work could sometimes be done over an Internet connection. But in conflicts where communications infrastructure is limited or overwhelmingly controlled by one side, proximity to the actors would be necessary, as would language skills and cultural knowledge.) Their stamp of approval would be a valuable commodity, a green light for media and other observers to take given content seriously. A state or warring party could bypass these monitors, but doing so would devalue whatever content was produced and make it highly suspect to others.

The monitors would examine the data, not the deed, so their conclusions would be weighted heavily, and states might launch inter-

---

* There is a start-up today called Storyful that does this for many of the major news broadcasters. It employs former journalists and carefully curates content from social media (e.g., by verifying that the weather in a YouTube video matches the weather recorded in that city on the day the video was supposedly shot).

ventions, send aid or issue sanctions based on what they say. And, of course, with such trust and responsibility comes the inevitable capacity for abuse, since these monitors would be no less immune to the corruption that stymies other international organizations. Regimes might attempt to co-opt verification monitors, through bribes or blackmail, and some monitors might harbor personal biases that reveal themselves too late. Regardless, the bulk of these monitors would be comprised of honest engineers and journalists working together, and their presence in a conflict would lead to more safety and transparency for all parties.

When not engaged in marketing wars, groups in conflict will attack whatever online entities they deem valuable to the other side. This means targeting the websites, platforms and communications infrastructure that have some strategic or symbolic importance with distributed denial of service (DDoS) attacks, sophisticated viruses and all other types of cyber warfare. Online attacks will become an integral part of the tactical strategy for groups in conflict, from the lowest intensity fight to full-fledged warfare. Attacking or incapacitating a rival group's communications network will not only interfere with its digital marketing abilities but will also affect its access to resources, information and its support base. Once a network or database has been successfully compromised, the infiltrating group can use the information they gathered to stay informed, spread misinformation, launch preemptive attacks and even track high-value targets (if, for example, a group found the mobile number for regime officials and had monitoring software that revealed their locations).

Virtual attacks will happen independently and in retaliation. In a civil war, for example, if one side loses territory to the other, it might retaliate by bringing down its rival's propaganda websites so as to limit its ability to brag about the victory—not an equivalent gain, of course, but damaging nonetheless. This is the virtual-world version of bombing the ministry of information, often one of the first targets in a physical-world conflict. A repressive government will be able to locate and disable the online financial portals that revolutionaries in the country are using to receive funds from supporters in the diaspora. Hackers sympathetic to one side or the other will take it upon themselves to dismantle whatever they can reach: YouTube channels run by their adversaries, databases relevant to the other side. When NATO began its military operations in Serbia in 1999, pro-Serbian hackers targeted public websites for both NATO and the U.S. Defense Department, with some success. (NATO's public-affairs website for Kosovo was "virtually inoperable" for days as a result of the attacks, which also seriously clogged the organization's e-mail server.)

In the coming decades, we'll see the world's first "smart" rebel movement. Certainly, they'll need guns and manpower to challenge the government, but rebels will be armed with technologies and priorities that dictate a new approach. Before even announcing their campaign, they could target the government's communications network, knowing it constitutes the real (if not official) backbone of the state's defense. They might covertly reach out to sympathetic governments to acquire the necessary technical components—worms, viruses, biometric information—to disable it, from within or without. A digital strike against the communications infrastructure would catch the government off guard, and as long as the rebels didn't "sign" their attack, the government would be left wondering where it came from and who was behind it. The rebels might leave false clues as to the origin, perhaps pointing to one of the state's external enemies, to confuse things further. As the state worked to get itself back online, the rebels might strike again, this time infiltrating the government's Internet and "spoofing" identities (tricking the network into believing the infiltrators are legitimate users) to further disorient and disrupt the network processes. (If the rebels gained access to an important biometric database, they could steal the identities of government officials and impersonate them online, making false statements or suspicious purchases.) Finally, the rebels could target something tangible, like the country's power grids, the manipulation of which would generate public outcry and blame, incorrectly aimed at the government. Thus the smart rebel movement could, with three digital strikes and no shots fired, find itself uniquely poised to mobilize the masses against a government that wasn't even aware of a domestic rebellion. At this point, the rebels could begin their military assault and open a second, physical front.

Conflicts in the future will also be influenced by two distinct and largely positive trends that stem from connectivity: first, the wisdom of the online crowd, and second, the permanence of data as evidence, which we alluded to earlier as making it harder for perpetrators of violence to deny or minimize their crimes.

Collective wisdom on the Internet is a controversial subject. Many decry the negative extremes of online collaboration, such as the aggressive mediocrity of the "hive mind" (the collective consensus of groups of online users) and the viciousness of anonymity-fueled pack behavior on forums, social networks and other online channels. Others champion the level of accuracy and reliability of crowd-sourced information platforms like Wikipedia. Whatever your view, there are potential gains that collective wisdom can bring to future conflict.

With a more level playing field for information in a conflict, a greater number of citizens can participate in shaping the narratives that emerge. Widespread mobile-phone usage will ensure that more people know what's going on inside a country than did in earlier times, and Internet connectivity extends that sphere of engagement to a broad range of outside actors. On balance, there are always more people on the side of good than on the side of the aggressors. With an engaged population, there is greater potential for citizen mobilization against injustice or propaganda: If enough people are angry with what they see, they'll have channels through which they can make their voices heard, and can act individually or collectively—even if, as we saw in Singapore, the anger is over the cooking of curry.

The challenges of governing the Internet also allow for the danger of online vigilantism, as the story of China's "human-flesh search engines" (*renrou sousuo yinqing*) shows. According to Tom Downey's revealing March 2010 article in *The New York Times Magazine,* some years ago a disturbing trend emerged in China's online space, where volumes of Internet users would locate, track down and harass individuals who had earned their collective wrath. (There is no central platform for this work, nor is the trend limited to China, but the phenomenon is most widely known and understood there, thanks to a series of high-profile examples.) In 2006, a gruesome video circulated on Chinese Internet forums depicting a woman stomping a kitten to death with her high-heeled shoes, leading to a countrywide search for the stomper. Through diligent crowd-sourced detective work, the perpetrator was soon tracked to a small town in northeastern China, and after her name, phone number and employer were made public, she fled, as did her cameraman. It's not just computers that can find needles in haystacks, apparently; locating this woman among more than one billion Chinese—through only the clues in the video—took just six days.

This kind of mob behavior can veer into unpredictable chaos, but that does not mean attempts to harness its collective power for good should be abandoned. Imagine if the end goal of the Chinese users was not to harass the kitten-stomper but to bring her to justice through official channels. In a conflict scenario, where institutions have broken down or are not trusted by the population, crowd-sourced energy will help to produce more comprehensive and accurate information, help track down wanted criminals and create demand for accountability even in the most difficult circumstances.

But the importance and utility of crowd-sourced justice pales in comparison to the other modern development: data permanence. The exposure of atrocities in real time and in front of a global audience is vital, as is permanently storing it and making it searchable for everyone who wants to refer to it (for prosecutions, legislation or later study). Governments and other aggressors may have the military advantage with guns, tanks and planes, but they'll be fighting an uphill battle against the information trail they leave behind. If a government attempts to block citizen communications, it may be able to stifle some of the evidence flowing through and out of the country, but the flow will continue. More important, the presence of this evidence, even if disputed at the time, will affect how the conflict is handled, resolved and considered well into the future.

Accountability, or the threat of it, is a powerful idea; that's why people try to destroy evidence. In the absence of hard data, conflicting narratives can impede justice and closure, and this applies to citizens and states alike. In January 2012, France and Turkey became embroiled in a diplomatic row when the French Senate passed a bill (struck down one month later by the French Constitutional Council) that made it illegal to deny that the mass killing of Armenians by the Ottoman Empire in 1915 was genocide. The Turkish government, which rejects the term "genocide" and claims that far fewer than 1.5 million Armenians were killed, called the bill "racist and discriminatory" and said judgment of the killings should be left to historians. With the technological devices, platforms and databases we have today, it will be much more difficult for governments in the future to argue over claims like these, not just because of permanent evidence but because everyone else will have access to the same source material.

In the future, tools like biometric data matching, SIM-card tracking and easy-to-use content-generating platforms will facilitate a level of accountability never before seen. A witness to a crime will be able to use his phone to capture what he sees and identify the perpetrator and victim with facial-recognition software almost instantly, without having to be directly in harm's way. Information about crimes or brutality in digital form will be automatically uploaded to the cloud (thus no data loss if the witness's phone is confiscated) and perhaps sent to an international monitoring or judicial body. An international court could then investigate, and depending on what it found, begin a public virtual trial and broadcast the proceedings back into the country where the perpetrator was roaming free. The risk of public shame and criminal charges might not deter leaders, but it would be enough to make some foot soldiers think twice before engaging in more violent activities. Professionally verified evidence would be available at The Hague's website before the trial, and witnesses would be able to testify virtually and in safety.

Of course, the wheels of justice turn slowly, particularly in the labyrinthine environment of international law. While a system of data responsiveness develops, the intermediate gains will be the storage of verifiable evidence, and better law enforcement will result. An open-source app, created by the International Criminal Court or some other body, could feature the world's most wanted criminals broken down by country. Just as the Chinese human-flesh search engines can pinpoint an individual's location and contact details, the same capability can be turned toward hunting down criminals. (Remember: People will have powerful phones in even the most remote places.) Using the same platform, concerned citizens around the world could contribute financially toward a reward as an incentive for making an arrest. Then, instead of facing mob justice, the criminal would be taken into custody by police and put on trial.

The collective power of the online world will serve as a tremendous deterrent to potential perpetrators of brutality, corrupt practices and even crimes against humanity. To be sure, there will always be truly malevolent types for whom deterrence will not work, but for merely dishonorable individuals, the potential costs of bad behavior in a digital age will become only greater. Beyond the heightened risks of accountability and the increased liklihood of a crime being documented and preserved in perpetuity, whistle-blowers will use technology to reach the widest possible audience. Defectors will have a far greater incentive to avoid accusations of complicity in these documented crimes as well. Perhaps a digital witness-protection program will be built to provide informants with new virtual identities (like the ones sold on the black market mentioned earlier) to offer further incentives for their participation.

Permanent digital evidence will also help shape transitional justice after a conflict has ended. Truth-and-reconciliation committees in the future will feature a trove of digital records, satellite surveillance, amateur videos and photos, autopsy reports and testimonials. (We'll explore this topic shortly.) Again, the fear of being held accountable will be a sufficient deterrent for some would-be aggressors; at the very least they might dial back the level of violence.

Beyond documenting atrocities, cloud storage will make data permanence relevant and important to people in conflict. Personal data not in the physical world will be safer, as it will be unreachable. Sometimes the outbreak of violence catches everyone by surprise. But in instances where the security situation is visibly declining, individuals will anticipate and prepare for the possibility of fleeing or being displaced. Individuals will also be able to sustain their claims to their homes, property and businesses even in exile or as refugees by capturing visual evidence and using tools like Google Maps and GPS to mark boundaries. They'll be able to move their land titles and deeds to the cloud. Where there are disputes, the digital platforms will assist in arbitration. Civilians caught up in conflict and forced to flee could take pictures of all of their possessions and re-create a model of their home in virtual space. If they return, they'll know exactly what is missing and may well be able to use a social-networking platform to locate the stolen items (after they've digitally verified that they own them).

Automated Warfare

When conflict escalates into armed combat, future participants will find the landscape of war to be nothing like it has been in the past. The opening of a virtual front to warfare will not change the fact that the most sophisticated automated weapons and soldiers must still operate in the physical world, never eliminating the essential role that human guidance and judgment play. But militaries that do not take into account this dual-world phenomenon (and their responsibilities in both) will find that, while new technology makes them far more efficient killing machines, they are hated and reviled as a result, making the problem of winning hearts and minds that much more difficult.

The modern automation of warfare, through developments in robotics, artificial intelligence and unmanned aerial vehicles (UAVs), constitutes the most significant shift in human combat since the invention of the gun. It is, as the military scholar Peter Singer notes in his masterly account of this trend, *Wired for War: The Robotics Revolution and Conflict in the 21st Century,* what scientists would call a "singularity"—a "state in which things become so radically different that the old rules break down and we know virtually nothing." Much as with other paradigm shifts in history (germ theory, the invention of the printing press, Einstein's theory of relativity), it is almost impossible to predict with any great accuracy how the eventual change to fully automated warfare will alter the course of human society. All we can do is consider the clues we see today, convey the thinking of people on the front lines, and make some educated guesses.

Integrating information technology into the mechanics of warfare is not a new trend: DARPA, the Pentagon's research-and-development arm, was created in 1958 as a response to the launch of Sputnik.* The government's determination to avoid being caught off guard again was such that DARPA's mission is, quite literally, "to maintain the technological superiority of the U.S. military and prevent technological surprise from harming our national security." Subsequently, the United States has led the world in sophisticated military technology, in everything from smart bombs to unmanned drones and bomb-

---

* Computer enthusiasts will remember this agency's central role in creating the Internet, back when the agency was known as Advanced Research Projects Agency (ARPA).

defusing explosive-ordnance-disposal (EOD) robots. But, as we'll discuss below, the United States may not hold that exclusive advantage for very long.

It's easy to understand why governments and militaries like robots and other unmanned systems for combat: They never tire, they never feel fear or emotion, they have superhuman capabilities and they always follow orders. As Singer points out, robots are uniquely suited to the roles that the military refers to as the three Ds (jobs that are dull, dirty or dangerous). The tactical advantages conferred by robots are constrained only by the limits of robotics manufacturers. They can build robots that withstand bullets, have perfect aim, recognize and disarm targets, and carry impossible loads in severe conditions of heat, cold or disorientation. Military robots have better endurance and faster reaction time than any soldier, and politicians will much more readily send them into battle than human troops. Most people agree that the introduction of robots into combat operations, whether on the ground, at sea or in the air, will ultimately produce fewer combat deaths, fewer civilian casualties and less collateral damage.

Already there are many forms of robots at work in American military operations. More than a decade ago, in 2002, iRobot, the company that invented the Roomba robotic vacuum cleaner, introduced a ground robot called the PackBot, a forty-two-pound machine with treads like a tank's, cameras and a degree of autonomous functionality, that military units can equip to detect mines, sense chemical or biological weapons and investigate potential IEDs (improvised explosive devices) along the sides of roads or anywhere else.* Another robotics manufacturer, Foster-Miller, makes a PackBot competitor called the TALON, as well as the first armed robot brought to battle: the Special Weapons Observation Reconnaissance Detection System, or SWORDS. And then there are the aerial drones. In addition to the now recognizable Predator drones, the U.S. military operates smaller versions (like the hand-launched Raven drone, used for surveillance) and larger ones (like the Reaper, which flies higher, faster and with a larger weapons payload than the Predator). An internal congressional

---

* Two PackBots were deployed during the Fukushima nuclear crisis following the 2011 earthquake in Japan, entering the damaged plant, where high radiation levels made it dangerous for human rescue workers, to gather visual and sensory data.

report acquired by *Wired* magazine's Danger Room blog in 2012 stated that drones now account for 31 percent of all military aircraft—up from 5 percent in 2005.

We spoke to a number of former and current Special Forces soldiers to gauge how they believed this progression of robotic technologies will affect combat operations in the next decades. Harry Wingo, a Googler and former Navy SEAL, spoke to the usefulness of using computers and "bots" instead of humans for surveillance, and of robots "taking point" in advancing through a field of fire or when clearing a building. In the next decade, he said, more "lethal kinetics"—operations involving fire—"will be handed over to bots, including those like room-clearing that require split-second parsing of targets." Initially, the robots will be operated with "machine-assist," meaning a soldier will direct the machine from a remote location, but eventually, Wingo believes, "the bots will identify and engage targets." Since 2007, the U.S. military has deployed armed SWORDS robots that can semi-autonomously recognize and shoot human targets, though it is believed that they have not, as yet, been used in a lethal context.

Soldiers will not be left behind completely, and not all human functions will be automated. None of the robots in operation today operate fully autonomously—which is to say, without any human direction—and, as we'll discuss later, there are important aspects of combat, like judgment, that robots will not be capable of exercising for many years to come. To better understand how technology will soon enhance the capabilities of human soldiers we asked a now inactive Navy SEAL, who, incidentally, participated in the Osama bin Laden raid in May 2011, what he anticipated for combat units in the future. First, he told us, he envisioned units equipped with highly sophisticated and secure tablet devices that will allow soldiers to tap into live video feeds from UAVs, download relevant intelligence analysis and maintain situational awareness of friendly troop movements. These devices will have unique live maps loaded with enough data about the surrounding environment—the historical significance of a street or building, the owners of every home, and the interior infrared heat movements captured by drones overhead—to provide soldiers with a much clearer sense of what to target and what to avoid.

Second, the clothes and gear that soldiers wear will change. Haptic technologies—this refers to touch and feeling—will produce uniforms that allow soldiers to communicate through pulses, sending out signals to one another that result in a light pinch or vibration in a particular part of the body. (For instance, a pinch on the right calf could indicate a helicopter is inbound.) Helmets will have better visibility and built-in communications, allowing commanders to see what the soldiers see and "backseat drive," directing the soldiers remotely from the base. Camouflage will allow soldiers to change their uniform's color, texture, pattern or scent. Uniforms might even be able to emit sounds to drown out noises soldiers want to hide—sounds of nature masking footsteps, for example. Lightweight and durable power sources will be integrated as well, so that none of the devices or wearable technologies will fail at crucial moments due to heat, water or distance from a charger. Soldiers will have the additional ability to destroy all of this technology remotely, so that capture or theft will not yield valuable intelligence secrets.

And, of course, wrapping all of this together will be a hefty layer of cybersecurity—more than any civilian would use—that enables instant data transmission within a cocoon of electronic protection. Without security, none of the advantages above will be worth the considerable cost that will be required to develop and deploy them.

Alas, military contractors' procedures will hold back many of these developments. In the United States, the military-industrial complex is working on some of the initiatives mentioned above—DARPA has led the development of many of the robots now in operation—but it is by nature ill-equipped to handle innovation. Even DARPA, while relatively well funded, is predictably stalled by elaborate contracting structures and its position in the Department of Defense bureaucracy. The innovative edge that is the hallmark of the American technology sector is largely walled off from the country's military by an anarchic and byzantine acquisitions system, and this represents a serious missed opportunity. Without reforms that allow military agencies and contractors to behave more like small private companies and start-ups (with maneuverability and the option to move quickly) the entire industry is likely to retrench rather than evolve in the face of fiscal austerity.

The military is well aware of the problems. As Singer told us, "It's a big strategic question for them: How do they break out of this broken structure?" Big defense projects languish in the prototype stage, over budget and behind schedule, while today's commercial technologies and products are conceived of, built and brought to market in volume in record time. For example, the Joint Tactical Radio System, which was supposed to be the military's new Internet-like radio-communications network, was conceived of in 1997, then shut down in September 2012, only to have acquisitions functions transferred to the Army under what is now called the Joint Tactical Networking Center. By the time it was shut down as its own operation, it had cost billions of dollars and was still not fully deployed on the battlefield. "They just can't afford that kind of process anymore," Singer said.

One recourse for the military and its contractors is to use commercial, off-the-shelf (COTS) products, which means buying commercially available technologies and devices rather than developing everything in-house. The integration of such outside products, however, is not an easy process; meeting military specifications alone (for ruggedness, utilization and security) can introduce damaging delays. According to Singer, the bureaucracy and inefficiency of the military contracting system have actually generated an unprecedented degree of ground-level ingenuity in building functional work-arounds. Some involve buying quick-need systems outside the normal Pentagon acquisitions process; that is how MRAP (mine-resistant, ambush-protected) vehicles were rushed to the front after the scourge of IEDs began in Iraq. And troops often adapt commercial technologies that they take on a deployment themselves.

Even military leaders have recognized the advantages that such inventiveness can bring. "The military was, in some ways, aided by the demands of the battlefields in Iraq and Afghanistan," Singer explained. "In Afghanistan, Marine attack helicopter pilots have taken to strapping iPads onto their knees as they fly, and using those for maps instead of the built-in system in their crafts."[*] He added that as the pressure of an active battlefield ends, military leaders are worried that innovative work-arounds might evaporate. It remains to be seen if innovation will drive change in a problematic contracting system.

. . .

---

[*] Singer's statement was corroborated by several active-duty Special Forces soldiers we spoke to.

Technological breakthroughs have offered the United States major strategic advantages in the past. For many years after the first laser-guided missiles were developed, no other country could match their lethality over long distances. But technological advantages generally tend to equalize over time, as technologies are spread, leaked or reverse-engineered, and sophisticated weaponry is no exception. The market for drones is already international: Israel has been at the forefront of that technology for years; China is very active in promoting and selling its drones; and Iran unveiled its first domestically built drone bomber in 2010. Even Venezuela has joined the club, utilizing its military alliance with Iran to create an "exclusively defensive" drone program that is operated by Iranian missile engineers. When asked to confirm reports of this program, the Venezuelan president Hugo Chavez remarked, "Of course we're doing it, and we have the right to. We are a free and independent country." Unmanned drones will get smaller, cheaper and more effective over time. As with most technologies, once a product is released into the environment—be it a drone or a desktop application—it's impossible to put it back in the box.

We asked the former DARPA director Regina Dugan how the United States approaches the high level of responsibility that comes with building such things, knowing that the ultimate consequences are out of its control. "Most advances in technology, particularly big ones, tend to make people nervous," she said. "And we have both good and bad examples of developing the societal, ethical and legal framework that goes with those kinds of technological advances." Dugan pointed to the initial concerns people expressed about human genome sequencing when that breakthrough was announced: If it could be determined that you had a predisposition toward Parkinson's disease, how would that affect how employers and insurance companies treated you? "What came to pass was the understanding that the advance that would allow you to see that predisposition was not the thing that we should shy away from," Dugan explained, "but rather we should create the legal protections that ensure that people couldn't be denied health care because they had a genetic predisposition." The development of technological advances and the protections they will ultimately require must grow in tandem for the right balance to be struck.

Dugan described her former agency's role in stark terms: "You can't undertake a mission like the invention and prevention of strategic surprise if you're unwilling to do things that initially make people feel uncomfortable." Rather, the obligation is to handle that job responsibly—which, critically, requires input and help from other people. "The agency can't do it by itself. One has to involve other branches of government, other parties, in the debate about those things," she said.

It is comforting to hear how seriously DARPA takes its responsibility for these new technologies, but the problem is, of course, that not all governments will approach them with similar consideration and caution. The proliferation of drones presents a particularly worrisome challenge, given the enormous benefits they bestow upon even the smallest armies. Not every government or military in the world has the technical infrastructure or human capital to support its own fleet of unmanned vehicles; only those with deep pockets will find it easy to buy that capability, openly or otherwise. Owning military robots—particularly unmanned aerial vehicles—will become a strategic prerogative for every country; some will acquire them to gain an edge, and the rest will acquire them just to maintain their sovereignty.

Underneath this state-level competition, there will be an ongoing race by civilians and other non-state actors to acquire or build drones and robots for their own purposes. Singer reminded us that "non-state actors that range from businesses like media groups and agricultural crop dusting to law enforcement, to even criminals and terrorists, have all used drones already." The controversial private military firm Blackwater, now called Academi, LLC, unveiled its own special service—unmanned drones, available to rent for surveillance and reconnaissance missions—in 2007. In 2009, it was contracted to load bombs onto CIA drones.

There is also plenty of private development and use of drones outside the context of military procurement. For example, some real-estate firms are now using private drones to take aerial photographs of their larger properties. Several universities have their own drones for research purposes; Kansas State University has established a degree for unmanned aviation. And in 2012 we learned about Tacocopter (a service allowing anyone craving a taco to order on a smart phone, punch in his location and receive his tacos by drone), which proved to be a hoax, but is both technically possible and not far off.

As we mentioned earlier, lightweight and inexpensive "everyman" drones engineered for combat purposes will become particularly popular at the global arms bazaar and in illicit markets. Remotely piloted model planes, cars and boats that can conduct surveillance, intercept hostile targets and carry and detonate bombs will pose serious challenges for soldiers in war zones, adding a whole other dimension to combat operations. If the civilian version of armed drones becomes sophisticated enough, we could well see military and civilian drones meeting in battle, perhaps in Mexico, where drug cartels have the will and the resources to acquire such weapons.

Governments will seek to restrict access to the key technologies making drones easy to mass-produce for the general populace, but regulating the proliferation and sale of these everyman drones will be very difficult. An outright ban is simply unrealistic, and even modest attempts to control civilian use in peaceful countries will have limited success. If, for example, the U.S. government required people to register their small unmanned aircrafts, restricted the spaces in which drones could fly (not near airports or high-value targets, for example) and banned their transport across state lines, it's not hard to imagine determined individuals finding ways around the rules by reconfiguring their devices, anonymizing them or building in some kind of stealth capacity. Still, we might see international treaties around the proliferation of these technologies, perhaps banning the sale of larger drones outside official state channels. Indeed, states with the greatest capacity to proliferate UAVs may even pursue the modern-day version of the Strategic Arms Limitation Talks (SALT), which sought to curtail the number of U.S. and Soviet arms during the Cold War.

States will have to work hard to maintain the security of their shores and borders from the growing threat of enemy UAVs, which, by design, are hard to detect. As autonomous navigation becomes possible, drones will become mini cruise missiles, which, once fired, cannot be stopped by interference. Enemy surveillance drones may be more palatable than drones carrying missiles, but both will be considered a threat since it won't be easy to tell the two apart. The most effective way to target an enemy drone might not be with brute force but electronically, by breaching the UAV's cybersecurity defenses. Warfare then becomes, as Singer put it, a "battle of persuasion"—a fight to co-opt and persuade these machines to do something other than their mission. In late 2011, Iran proudly displayed a downed but intact American drone, the RQ-170 Sentinel, which it claimed to have captured by hacking into its defenses after detecting it in Iranian airspace. (The United States, for its part, would say only that the drone had been "lost.") An unnamed Iranian engineer told *The Christian Science Monitor* that he and his colleagues were able to make the drone "land on its own where we wanted it to, without having to crack the remote-control signals and communications" from the U.S. control center because of a known vulnerability in the plane's GPS navigation. The technique of implanting new coordinates, known as spoofing, while not impossible, is incredibly difficult (the Iranians would have had to get past the military's encryption to reach the GPS, by spoofing the signals and jamming the communications channels).

Diplomatic solutions might involve good-faith treaties between states not to send surveillance drones into each other's airspace or implicit agreements that surveillance drones are an acceptable offense. It's hard to say. Perhaps there might emerge international requirements that surveillance drones be easily distinguishable from bomber drones. Some states might join together in a sort of "drone shield," not unlike the nuclear alliance of the Cold War, in which case we would see the world's first drone-based no-fly zone. If a small and poor country cannot afford to build or buy its own bomber drones, yet it fears aerial attacks from an aggressive neighbor, it might seek an alliance with a superpower to guarantee some measure of protection. It seems unlikely, however, that states without drones will remain bereft for long: The Sentinel spy drone held by the Iranians cost only around $6 million to make.

The proliferation of robots and UAVs will increase conflict around the world—whenever states acquire them, they'll be eager to test out their new tools—but it will decrease the likelihood of all-out war. There are a few reasons for this. For one, the phenomenon is still too new; the international treaties around weapons and warfare—the Nuclear Nonproliferation Treaty, the Anti-Ballistic Missile Treaty, and the Chemical Weapons Convention, to name a few—have not caught up to the age of drones. Boundaries need to be drawn, legal frameworks need to be developed and politicians must learn how to use these tools responsibly and strategically. There are serious ethical considerations that will be aired in public discourse (as is taking place in the United States currently). These important issues will lead states to exhibit caution in the early years of drone proliferation.

We must also consider the possibility of a problem with loose drones, similar to what we see with nuclear weapons today. In a country such as Pakistan, for example, there are real concerns about the state's capacity to safeguard its nuclear stockpiles (estimated to be a hundred nuclear weapons) from theft. As states develop large fleets of drones, there will be a greater risk that one of these could fall into the wrong hands and be used against a foreign embassy, military base or cultural center. Imagine a future 9/11 committed not by hijackers on commercial airliners, but instead by drones that have fallen out of state hands. These fears are sufficient to spur future treaties focused on establishing requirements for drone protection and safeguarding.

States will have to determine, separately or together, what the rules around UAVs will be, whether they will be subject to the same rules as regular planes regarding violating sovereign airspace. States' mutual fears will guard against a rapid escalation of drone warfare. Even when it was revealed that the American Sentinel drone had violated Iranian airspace, the reaction in Tehran was boasting and display, not retaliation.

The public will react favorably to the reduced lethality of drone warfare, and that will forestall outright war in the future. We already have a few years of drone-related news cycles in America from which to learn. Just months before the 2012 presidential election, government leaks resulted in detailed articles about President Obama's secret drone operations. Judging by the reaction to drone strikes in both official combat theaters and unofficial ones like Somalia, Yemen and Pakistan, lethal missions conducted by drones are far more palatable to the American public than those carried out by troops, generating fewer questions and less outrage. Some of the people who advocate a reduced American footprint overseas even support the expansion of the drone program as a legitimate way to accomplish it.

We do not yet understand the consequences—political, cultural and psychological—of our newfound ability to exploit physical and emotional distance and truly "dehumanize" war to such a degree. Remote warfare is taking place more than at any other time in history and it is only going to become a more prominent feature of conflict. Historically, remote warfare has been thought of mostly in terms of weapons delivered via missiles, but in the future it will be both commonplace and acceptable to further separate the actor from the scene of battle. Judging from current trends, we can assume that one effect of these changes will be less public involvement on the emotional and political levels. After all, casualties on the *other* side are rarely the driving factor behind foreign policy or public sentiment; if American troops are not seen to be in harm's way, the public interest level drops dramatically. This, in turn, means a more muted population on matters of national security; both hawks and doves become quieter with a smaller threat to their own soldiers on the horizon. With more combat options that do not inflame public opinion, the government can pursue its security objectives without having to consider declaring war or committing troops, decreasing the possibility of outright war.

The forecast of fewer civilian casualties, less collateral damage and the reduced risk of human injury are welcome, but the shift toward a more automated battlefield will introduce significant new vulnerabilities and challenges. Chief among them will be maintaining the cybersecurity of equipment and systems. The data flow between devices, ground robots and UAVs, and their human-directed command-and-control centers must be fast, secure and unimpeded by poor infrastructure, just like communications between troop units and their bases. This is why militaries set up their own communications networks instead of relying on the local one. Until robots in the field have autonomous artificial intelligence, an impeded or broken connection turns these machines into expensive dead weight—possibly dangerous, too, since capture of an enemy's robot is akin to capturing proprietary technology. There is no end to the insights such a capture could yield, particularly if the robot is poorly designed—not only information about software and drone engineering, but even more sensitive data like enemy locations gleaned through digital coordinates. (It's also hard to imagine that countries won't purposely crash-land or compromise a decoy UAV, filled with false information and misleading technical components, as part of a misinformation campaign.) In wars where robotic elements are present, both sides will employ cyber attacks to interrupt enemy activity, whether by spoofing (impersonating a network identity) or employing decoys to disrupt enemy sensor grids and degrade enemy battle networks. Manufacturers will attempt to build in fail-safe mechanisms to limit the damage of these attacks, but it will be difficult to build anything technologically bulletproof.

Militaries and robotics developers will face simple error as well. All networked systems have vulnerabilities and bugs, and often the only way they become known is when they are revealed by hackers or independent security-systems experts. The computer code necessary to operate machines of this caliber is incredibly dense—millions upon millions of lines of code—and mistakes happen. Even when developers are aware of a system's vulnerabilities, it isn't easy to address them. The vulnerability the Iranians said they attacked in bringing down the U.S. drone, a weakness in the GPS system, had reportedly been known to the Pentagon since the Bosnian campaign of the 1990s. In 2008, U.S. military troops first discovered laptops from Shiite insurgents in Iraq containing files of intercepted drone video feeds, which the Iraqis had been able to access by simply pointing their satellite dishes up and using a cheap downloadable software, SkyGrabber, available for $26, that was originally intended to help people pirate movies and music. The data links between the drone and its ground control station were never encrypted.

For the near future, as humans continue to drive the implementation of these technologies, mistakes will be made. Placing fragile human psyches in extreme combat situations will always generate unpredictability—and can trigger PTSD, severe emotional distress or full psychotic breaks in the process. As long as human beings conduct war, these errors must be factored in.

Until artificially intelligent systems can mimic the capability of the human brain, we won't see unmanned systems entirely replacing human soldiers, in person or as decision-makers. Even highly intelligent machines can have glaring faults. As Peter Singer pointed out, during World War I, when the tank first appeared on the battlefield, with its guns, armor and rugged treads, it was thought to be indestructible—until someone came up with the antitank ditch. Afghanistan's former minister of defense Abdul Rahim Wardak, whom we met in Kabul shortly before he was dismissed, chuckled as he described how he and his fellow mujahideen fighters targeted Soviet tanks in the 1980s by smearing mud on their windows and building leaf-covered traps similar to the ones the Vietcong used to ensnare American soldiers a decade earlier. In a modern parallel, Singer said, "The ground robots our soldiers use in Iraq and Afghanistan [employ] an amazing technology, but insurgents realized they could build tiger traps for them—just deep holes that [they] would fall into. They even figured out the angle necessary for the incline so that the bot couldn't climb its way out." The intelligence of these robots is specialized, so as they are tested in the field, their operators and developers will continually encounter enemy circumventions that *they* did not expect, and they'll be forced to evolve their products. Asymmetric encounters in combat like these will continue to pose unpredictable challenges for even the most sophisticated of technologies.

Human intelligence contains more than just problem-solving skills, however. There are uniquely human traits relevant to combat—like judgment, empathy and trust—that are difficult to define, let alone instill in a robot. So what is lost as robots increasingly take over human responsibilities in battlefield operations? In our conversations with Special Forces members, they emphasized the supreme importance of trust and brotherhood in their experiences in combat. Some had trained and fought together for years, coming to know each other's habits, movements and thought patterns almost instinctively. They described being able to communicate with just a look. Will robots ever be able to mimic a human's ability to read nonverbal cues?

Can a robot be brave? Can it selflessly sacrifice? Can a robot, trained to identify and engage targets, have some sense of ethics or restraint? Will a robot ever be able to distinguish between a child and a small man? If a robot kills an innocent civilian, who is to be blamed? Imagine a standoff between an armed ground robot and a six-year-old child with a spray-paint canister, perhaps sent out by an insurgent group. Whether acting autonomously or with human direction, the robot can either shoot the unarmed child, or be disabled, as the six-year-old sprays paint over its high-tech cameras and sensory components, blinding it. Faced with this decision, if you were commanding the robot, Singer asks, what would you do? We can't court-martial robots, hold them accountable or investigate them. Accordingly, humans will continue to dominate combat operations for many years to come, even as robots become more intelligent and integrated with human forces.

New Interventions

The advent of virtualized conflict and automated warfare will mean that states with aggressive agendas will have a wider range of tools available to them in the future. Interventions by other actors—citizens, businesses and governments—will diversify as well.

For states, the U.N. Security Council will remain the only international body that is both inclusive of all nations and capable of bestowing legality to state-led military interventions. It's unlikely that the international community will stray far from the great power dispensation of 1945 that established the United Nations, even with the vociferous calls of empowered citizen populations increasing the pressure on states to act. New mandates and charters for intervention will be almost impossible to pass given the fact that any amendment to the U.N. charter requires 194 member-nations to approve.

But there are areas of high-level statecraft where new forms of intervention are more viable, and these will take place through smaller alliances. In an extreme situation, we foresee a group of countries, for example, coming together to disable an errant country's military robots. We can also imagine some member-states of NATO pushing to establish new mandates for intervention that could authorize states to send combat troops into conflicts to establish safe zones with independent and uncompromised networks. This would be a popular idea within intervention policy circles—it's a natural extension of the Responsibility to Protect (RtoP) doctrine, which the U.N. Security Council used to authorize military action (including air strikes) in Libya in 2011 that NATO subsequently carried out. It's very possible that we will see NATO members contribute drones to enforce the world's first unmanned no-fly zone over a future rebel stronghold, which would not involve sending any troops into harm's way.

Beyond formal institutions like NATO, the pressure for action will find other outlets in the form of ad hoc coalitions involving citizens and companies. Neither individuals nor businesses are able to muster military force for a ground invasion, but they can contribute to the maintenance of the vitally important communications network in a conflict zone. Future interventions will take the form of reconnecting the Internet or helping a rebel-held area set up an independent and secure network. In the event of state or state-sponsored manipulation of communications, we'll see a concerted effort by international stake-holders to intervene and restore free and uninterrupted access without waiting for U.N. approval.

It's not the connectivity that is crucial per se (civilians in conflict zones might already have some form of communications access) but rather what a secure and fast network enables people to do. Doctors in makeshift field hospitals will be able to coordinate quickly, internally and internationally, to distribute medical supplies, arrange airdrops and document what they're seeing. Rebel fighters will communicate securely, off the government's telecommunications network, at ranges and on platforms much more useful than radios. Civilians will interact with members of their families in the diaspora on otherwise blocked platforms and use safe channels—mainly an array of proxy and cir-cumvention tools—to send money in or information out.

Coalitions of states could send the equivalent of Special Forces troops to help rebel movements disconnect from the government net-work and establish their own network. Today, actions like these are taken but in independent fashion. A group of Libyan ministers told us the story of a brave American soul called Fred who arrived in the rebel stronghold of Benghazi in a wooden boat, armed with commu-nications supplies and determined to help the rebels build their own telecommunications network. Fred eliminated the Gadhafi-era wire-taps as his first task. In the future, this will be a combat operation, particularly in places not accessible from the sea.

The composition of intervening coalitions will change in turn. States with small militaries but strong technology sectors will become new power players. Today, Bangladesh is among the most frequent contrib-utors of troops to international peacekeeping missions. In the future, it will be countries with strong technology sectors, presently including Estonia, Sweden, Finland, Norway and Chile, who lead the charge in this type of mission. Coalitions of the connected will bring the politi-cal will and digital weaponry like high bandwidth, jerry-rigged inde-pendent mobile networks and enhanced cybersecurity. Such countries might also contribute to military interventions, with their own robot and aerial-drone armies. Some states, particularly small ones, will find it easier, cheaper and more politically expedient to build and commit their own unmanned drone arsenal to multilateral efforts, rather than cultivating and deploying human troops.

Technology companies, NGOs and individuals will also participate in these coalitions, each bringing something uniquely valuable to the table. Companies can build open-source software tailored to the needs of the people inside a country, and offer free upgrades for all of their products. NGOs can coordinate with telecoms to build accurate databases of a given population and its needs, mapping out where the most unstable or isolated pockets are. And citizens can volunteer to test the new network and all of these products, helping to find bugs and vulnerabilities as well as providing crucial user feedback.

No matter how advanced our technology becomes, conflict and war will always find their roots in the physical world, where the decisions to deploy machines and cyber tactics are fundamentally human. As an equal-opportunity enabler, technology will enhance the abilities of all participants in a conflict to do more, which means more messaging and content from all sides, greater use of robots and cyber weapons, and a wider range of strategic targets to strike. There are some distinct improvements, like the accountability driven by the permanence of evidence, but ultimately technology will complicate conflict even as it reduces risk on a net level.

Future combatants—states, rebels, militaries—will find that the tough ethical, tactical and strategic calculations they are used to making in physical conflicts will need to account for a virtual front that will oftentimes affect their decision-making. This will lead aggressors to take more actions in the less risky virtual front, as we described earlier, with online discrimination and hard-to-attribute cyber first-strike invasions. In other instances, the virtual front will act as a constraining force, leading aggressors to second-guess the degree of their aggression on the physical front. And as we will see even more clearly in the following pages, the mere existence of a virtual front paves the way for intervention options that are still robust, but minimize or reduce altogether the need to send troops into harm's way. Drone-patrolled no-fly zones and robotic peacekeeping interventions may be possible during a conflict, but such steps are limited. When the conflict is over, however, and the reconstruction effort begins, the opportunities for technology to help rebuild the country are endless.

CHAPTER 7

The Future of Reconstruction

It's now eminently clear how technology can be used to turn societies upside down and even tear them apart, but what about putting them back together? Reconstruction after a conflict or a natural disaster is a long and arduous process, hardly something a flash mob or viral video campaign can carry out. But while communication technologies alone can't rebuild broken societies, political, economic and security efforts can all be enhanced and accelerated because of technology. Tools that we use for casual entertainment today will find new purpose in the future in postcrisis countries, and populations in need will find more information and more power at their fingertips. Reconstruction efforts will become more innovative, more inclusive and more efficient over time, as old models and methods are either updated or discarded. Technology cannot thwart disaster or halt a civil war, but it can make the process of putting the pieces back together less painful.

Just as future conflicts will see the addition of a virtual front, so too will reconstruction efforts. We will still see cranes and bulldozers restoring roads, rebuilding bridges and resurrecting destroyed buildings, but we will also see an immediate and simultaneous focus on key functions that in the past have often come later in the process. Getting communications up and running, for example, will enable the rebuilding of the physical infrastructure and the economic and governance infrastructure at the same time. Here we will outline how we envision the approach future reconstruction planners will take to a postcrisis society, discuss the wave of new participants that connectivity will spur to action and offer a few ideas for innovative policies that can put societies on a faster path toward recovery.

Communications First

For societies emerging from a man-made or natural disaster, reconstruction is a daunting task. From rebuilding roads and buildings to reconnecting the population to the services it needs, these challenges require immense resources, different types of technical expertise and, of course, patience. Modern technology can aid these processes significantly if employed in the right ways, and we believe that successful reconstruction efforts in the future will rely heavily on communication technologies and fast telecommunications networks.

There will be a reconstruction prototype: a flexible and segmentary set of adaptable practices and models that can be tailored to fit particular postcrisis environments. Technology companies use prototypes and "beta" models to allow room for trial and error—the underlying philosophy being that early-stage feedback for an imperfect product ultimately yields a better result in the end. (Hence the tech entrepreneur's favorite aphorism: Fail early, fail often.) A prototype-like approach to reconstruction efforts will take some time to develop, but ultimately it will better serve the communities in need.

The main component of a reconstruction prototype—and what distinguishes it from, say, more traditional reconstruction efforts—is a communications-first, or mobile-first, mentality. The restoration and upgrading of communication networks have already become the new cement in modern reconstruction efforts. Looking ahead, upgrading broken societies to the fastest and most modern version of telecommunications infrastructure will be the top priority of all reconstruction actors, not least because the success of their own work will depend on it. Even in the last decade we've witnessed such a shift.

As recently as the early 2000s, post-conflict reconstruction wasn't so much about telecommunications revival as it was telecommunications installation. Neither Afghanistan nor Iraq had any semblance of a mobile network prior to regime change. The Taliban government violently opposed almost every form of consumer technology (although it had a small GSM [Global System for Mobile Communications] network limited to government officials) and Saddam Hussein banned mobile phones entirely in his totalitarian state. Once those regimes fell, the populations were left with virtually no infrastructure or modern devices; combatants in the ensuing conflicts were the only ones with some form of portable communications (typically radios).

When American civilian reconstruction teams entered Iraq in 2003, they found themselves in a telecommunications desert, and initial efforts to use satellite phones floundered as they discovered that the phones worked only if both users stood outside—needless to say, an inconvenient feature for a war zone.* As a quick fix, the allies' Coalition Provisional Authority (CPA) gave MTC-Vodafone, a regional telecom company, a contract to install cell towers and establish services in the south of the country, while another telecom, MCI, got the nod in Baghdad. According to one former senior CPA official we spoke with, the towers were put up all over the country literally overnight, with officials and U.N. staff receiving thousands of mobile phones to distribute to important local political players. (Oddly enough, all the phones sported a "917" area code, sharing that distinction with New York's five boroughs.) These efforts jump-started a moribund telecommunications industry in Iraq by building the physical infrastructure required, and within a few years, the sector was booming.

In Afghanistan, where the U.N. established a mobile network soon after the fall of the Taliban (with free service as an incentive for users), the mobile market has grown significantly in the past decade, thanks largely to the Afghan government's decision to issue licenses to private mobile operators. By 2011, there were four major operators in Afghanistan, claiming some 15 million subscribers among them. The reconstruction teams who arrived in Iraq and Afghanistan found a blank canvas: poor infrastructure, no subscribers and dubious commercial prospects. Given the rate of mobile adoption around the world and how the telecommunications industry is expanding, it's unlikely that anyone will ever encounter a similar blank slate again.

In Haiti after the 2010 earthquake, the primary communications task was not installation but widespread restoration of a badly dam-

---

* These difficulties were compounded by the fact that the United States set up operational headquarters in Saddam Hussein's former palaces, which had been turned into electronically shielded bunkers by the paranoid dictator.

aged telecommunications infrastructure. Despite the devastation throughout the country, getting its communications networks up and running was a relatively fast process. The mobile infrastructure was badly damaged by the earthquake and aftershocks, but due to quick thinking and cooperation between local telecoms and the U.S. military, the carriers were able to restore functionality within only a few days. Ten days after the earthquake, the two largest mobile phone operators, Digicel and Voilà, reported that they were able to operate at 70 to 80 percent of their pre-earthquake capacity.

Jared, who was then with the State Department, remembers reaching out to the U.S. ambassador to Indonesia shortly after the Haitian earthquake for a debriefing on lessons learned after the 2004 tsunami that killed 230,000 people in fourteen countries in Southeast Asia. The message was clear: Get the towers up, get them running and overrule the people who think that telecommunications are secondary to emergency rescue. Fast networks aren't secondary; they're complementary.

Because the vast majority of cell towers in Haiti, even prior to the earthquake, relied on generators instead of electricity for power, maintaining coverage was often more a question of fuel than infrastructure. Donated cell towers had to be guarded lest desperate people try to steal their fuel. Still, the ability to maintain service despite the destruction and chaos proved vital in coordinating and sending aid organizations to areas and people who needed help most, as well as providing a way for friends and family to contact each other within and beyond Haiti. Some of the first images to come out of the country after the disaster were indeed taken and sent by Haitians using their mobile phones. Everyone involved in the immediate aftermath of the earthquake recognized how crucial working communications were in the midst of widespread physical destruction and human suffering.

The uprisings in the Arab world that began in 2010 represent another recent example of the advantages of a communications-first perspective. Vodafone's speedy restoration of service in Egypt just before Hosni Mubarak stepped down as president foreshadows a more agile and shrewd telecom sector. Vodafone's Vittorio Colao told us, "We had people sleeping in the network centers in order to make sure that we could be the first to offer service once the shutdown ended. We had food and water; we'd rented rooms in nearby hotels and we protected our premises, to make sure nobody could come and [disable] the network." As a result of its efforts, Vodafone was the first operator to resume service—an important "first" for a company trying to reach a large Egyptian market that suddenly had a lot to talk about. Colao described a smart and empathetic strategy on the part of Vodafone to demonstrate value to its Egyptian customers: "We gave credit to our Egyptian customers so that they could call people at home, as a give-away." Vodafone also shaped the traffic load (that is, freed up space on the network for Egyptian users), "so that when the network came back up, we could make sure the first people using it could [make] twenty euros' worth of calls to let relatives know [they were] safe."

Today's reliance on telecommunications is a reflection of how important this technology has become in even the poorest societies. In most cases today, when we talk about restoring the network, we're specifically talking about voice and text services—not Internet connectivity. This will change in the next decade, as people everywhere begin to rely more on data services than on voice communications. After a crisis, the pressures to restore Internet connectivity will dwarf what we see today with voice and text, both for the sake of the population and because a fast data network will help reconstruction actors achieve their goals. If necessary, aid organizations will deploy portable 4G towers meshed together into a low-bandwidth ISP. Data can hop from a mobile device to the nearest tower, then from tower to tower until it reaches a fiber-optic cable connecting to the broader Internet. Browsing speeds will be slow, but such portable deployments will provide enough connectivity to accelerate rebuilding.

Dedicated leadership by the telecommunications industry will be a feature of the reconstruction prototype, with telecoms leading the way as nationalized entities or coalition partners if they are in the private sector. Today, Bechtel and other engineering corporations are often tasked with rebuilding physical infrastructure through government contracts, but as the world adopts a communications-first outlook, the telecoms will be first in—and, like others, they'll come to make money. In postcrisis societies, solid networks are needed as soon as possible to coordinate search-and-rescue efforts; engage with the population; preserve the rule of law; organize and facilitate aid-distribution efforts; locate missing people; and help those who have been internally displaced navigate their new environment. Telecom companies will have clear and valid commercial motivations to invest their resources in building and maintaining a modern communications network. If the telecom sector is properly regulated from the beginning, the collective benefit for all parties will be quite high: The companies will earn revenue, the reconstruction actors will have faster and better tools, and the population at large will be able to access service that is reliable, fast and cheap (particularly if the sector is competitive from the outset).

The long-term benefit of a healthy telecommunications sector is that it promotes and facilitates the growth of the economy, even if the stability is slow to return. In general, direct investments in infrastructure, jobs and services offer more to the economy than short-term aid programs, and telecommunications is among the most universally lucrative and sustainable enterprises in the commercial world. Afghanistan's largest mobile operator, Roshan, is also the country's biggest investor and taxpayer. Roshan employs thousands in Afghanistan and provides nearly 5 percent of the Afghan government's overall domestic revenue. This is true despite substandard infrastructure, low incomes and more than a decade of continuous war. In the future, smart actors in reconstruction efforts—governments, multinational organizations and aid groups—will recognize the telecoms' value immediately and prioritize network building accordingly, rather than considering telecoms to be competitors or afterthoughts.

Because telecommunications is a profitable business (and never more so than after a crisis, when activity levels are unusually high), there will be ample opportunity for local and transnational entrepreneurs to participate. Talented local engineers will use open-source software to build their own platforms and applications to help the nascent economy, or they will collaborate with outside companies or organizations and contribute their skills. Much of the investment in the telecommunications space will be straightforward transactions and efforts to provide helpful services to the population, but there is some risk that the business leaders who emerge will come to constitute a new digital oligarchy. They might be well-connected local businessmen, taking advantage of the post-disaster environment to capture a key industry, or foreign executives looking to expand their empire. Regulation, again, will be key: As with all reconstruction efforts, those in charge will have to be wary of such maneuvers in a chaotic and highly malleable environment, and use their oversight effectively.

Mixed with entrepreneurs and digital oligarchs will be another group of foreign investors, members of the country's diaspora and others whose interests are personal rather than simply financial. In the future, investors looking to connect with new countries will find that global connectivity produces a much deeper and more multifaceted type of engagement. Real-time news alerts, active social networks and instant language translation will enable investors to feel much closer to the countries they operate in, akin to the deep knowledge possessed by diasporic communities around the world. This will lead to better and longer investments and a more fruitful relationship for both the investors and the societies with which they interact.

Few understand this better than Carlos Slim Helú, the Mexican telecom magnate and currently the world's richest person. Slim is also a part of the fifteen-million-strong Lebanese diaspora—his father emigrated to Mexico from Lebanon in 1902, fleeing the conscription of the Ottoman Empire army. Today, through a variety of companies, Slim maintains business interests around the world (including an 8 percent stake in *The New York Times*). He described to us how his experience as a child of immigrants has shaped his perspective. "I think that more than feeling just Lebanese, I feel I am part of the world altogether," he said. "Today, I feel I am a compromise between being Lebanese and relating to the challenges there, but also being a businessman in Latin America and with the responsibility I feel towards countries where I am doing business."

His experience is not unique, he explained, and in the future he predicted that everyone will become "more global *and* more local," with overlapping regional interests born from personal heritage, business opportunities and plain curiosity. He described himself as part of a new group that he calls the "business diaspora," where, as a transnational businessman, he believes, "We are not going to countries just to put money in and pull it out. We are making business to stay and be part of the development of the country." You can look at this as something "romantic," he added, but it's also smart business: "The reality is that business gets better if you grow the market, the demand, the customers and the possibilities."

As entry barriers lower for business in an increasingly interconnected world, the experience of being a member of the "business dias-pora" will not just be reserved for those with the means to invest large amounts of capital. Imagine, for example, that a computer-science student in Indiana develops a game for a popular social-networking site that suddenly takes off among users in Sri Lanka. The student and aspiring entrepreneur might not even have a passport (much less know anything about Sri Lanka), yet his game becomes highly profit-able, whatever the reasons. His curiosity piqued, the student adds Sri Lankan friends on Facebook and Google+, follows local news on Twit-ter and begins to learn about, and travel to, the country. In short order, he develops a digital kinship with the country, which will last for years to come. Millions of entrepreneurs, apps developers and businessmen will experience something similar in the future, because the markets online will be bigger and more diverse than anyone truly anticipates.

In a reconstruction setting, this outlook is of course encouraging, but even the most organized and well-meaning telecom companies will never supplant the heavy-duty work of governing institutions. There are basic goods and social services that only a government can provide to its population, like security, public-health programs, clean water supplies, transport infrastructure and basic education. Connectivity and telecommunications will improve the efficacy of these functions but only in partnership with institutional actors on the ground, as the following example shows.

With its initial collapse, in 1991, Somalia became the world's pre-mier failed state. Famines, clan warfare, external aggression, terrorist insurgencies and regional fragmentation have foiled transitional gov-ernment after transitional government. Over the past several years, the growth of mobile phones in Somalia has been one of the few success stories to emerge amid this anarchy. Even in the absence of security or a functioning government, the telecommunications industry has come to play a critical role in many aspects of society, providing Somalis with jobs, information, security and critical connections to the outside world. In fact, the telecoms are just about the only thing in Somalia that is organized, that transcends clan and tribal dynamics, and that functions across all three regions: South Central Somalia (Mogadishu), Puntland in the northeast and Somaliland in the northwest. Only one commercial bank exists in Somalia (founded in May 2012), and until there were mobile phones, in order to move money Somalis had to rely on informal *hawala* networks, in which no transaction records are kept. Today, mobile money-transfer services allow hundreds of thousands of Somalis to move money around inside the country and receive remittances from abroad. SMS-based platforms allow subscribers to use e-mail and receive stock tips and weather information.

Foreign NGOs and companies regularly launch mobile-technology pilot projects to improve the prospects for the Somali population in small ways; we've seen attempts to build SMS-based job-matching platforms and remote-diagnosis mobile health-care systems, among others. Yet most are unsuccessful in establishing a foothold—unsurprising, given the exceptionally hostile security and business environment. So most of the innovation that comes from Somalia today comes from the Somalis themselves; in this as elsewhere in the developing world, the most creative solutions emerge at the local level, driven by necessity more than anything else.

The absence of government in Somalia has meant that the telecommunications sector is unregulated, which drives down prices because entrepreneurs can step in and build a network if they see an opportunity (and have a sufficient appetite for risk). This is a common pattern when a government stops functioning. In the weeks after the fall of Saddam Hussein, a Bahraini telecom tried to expand into southern Iraq and capitalize on sectarian ties between that area, which is largely Shia, and Bahrain, which has a Shia majority, to win new customers. The occupying military forces, concerned about inflaming sectarian tensions, ultimately blocked the telecom's venture.

The extreme laissez-faire business environment in Somalia has produced some of the cheapest local, international and Internet rates in Africa, making mobile usage far more possible for a deeply impoverished population. When members of the Somali diaspora in the United States call their family back home, their relatives will often hang up and call *them* back. Without a government demanding taxes, charging for licenses or imposing regulatory costs, telecoms can keep costs low to expand their subscriber base while still turning a profit. Somalia's mobile penetration is much higher than one might expect, hovering somewhere between 20 and 25 percent. The four main telecom operators offer voice and data service across the country, and sixty to seventy miles into neighboring Kenya as well.

Despite these achievements in communications, Somalia remains an exceptionally insecure country, and insurgents have used the country's connectivity to further this volatility. Al-Shabaab Islamist insurgents send threatening calls and messages to African Union peacekeepers. Islamist radicals impose bans on mobile banking platforms and sabotage telecommunications infrastructure. Pirates on the Somali coast use local telecom networks to communicate because they worry their satellite phones can be tracked by international warships. In a February 2012 report, the United Nations Security Council added the head of Somalia's largest telecom, Hormuud, to its list of individuals subject to a travel ban after identifying him as one of al-Shabaab's chief financiers. (The report also said the man, Ali Ahmed Nur Jim'ale, set up Hormuud's mobile money-transfer system in order to facilitate anonymous funding to al-Shabaab.)

Certainly, the situation in Somalia is complex. But should the country emerge from its cocoon of instability anytime soon, the new government will surely find willing partners in the national telecom operators.

Ideally, reconstruction efforts strive not only to re-create what existed before, but to improve on the original and develop practices and institutions that reduce the risk of repeated disasters. The majority of postcrisis societies, while diverse in detail, have the same basic needs, roughly analogous to the basic components of state-building. These include administrative control of territory, a monopoly on the means of violence, sound management of public finances, investment in human capital, ensuring the provision of infrastructure and creating citizenship rights and duties.* Efforts to meet these needs, while heavily dependent on the international community (financially, technically and diplomatically), must be led by the postcrisis state itself. If reconstruction is not seen as homegrown or at least consistent with the political and economic aims of the society, the likelihood of failure increases dramatically.

Technology will help protect property rights, safeguarding virtual

---

* We take these duties from a list of the ten functions of the state in the book *Fixing Failed States,* by Clare Lockhart and Ashraf Ghani, the founders of the Institute for State Effectiveness.

records of real assets so that those assets can be quickly reclaimed when stability returns. Investors are not likely to put their money into a country where they feel insecure about the safety and ownership of their property. In post-invasion Iraq, three commissions were created to allow local people and returning exiles to reclaim or receive compensation for property seized during Saddam Hussein's regime. A parallel authority was set up to resolve disputes. These were important steps in the reconstruction of Iraq, serving as a moderating factor to the exploitation of post-conflict instability and instances of claiming property by force. But despite their good intentions (more than 160,000 claims were received by 2011), these commissions were hampered by certain bureaucratic restrictions that trapped many claims in complicated litigation. In the future, states will learn from this Iraqi model that a more transparent and secure form of protection for property rights can forestall such hassles in the event of conflict. By creating online cadastral systems (i.e., online records systems of land values and boundaries) with mobile-enabled mapping software, governments will make it possible for citizens to visualize all public and private land and even submit minor disputes, like a fence boundary, to a sanctioned online arbiter.

In the future, people won't just back up their data; they'll back up their government. In the emerging reconstruction prototype, virtual institutions will exist in parallel with their physical counterparts and serve as a backup in times of need. Instead of having a physical building for a ministry, where all records are kept and services rendered, that information will be digitized and stored in the cloud, and many government functions will be conducted on online platforms. If a tsunami destroys a city, all ministries will continue to function with some competence virtually while they are reconstituted physically.

Virtual institutions will allow new or shell-shocked governments to maintain much of their effectiveness in the delivery of services, as well as keep those governments an integral part of all reconstruction efforts. Virtual institutions won't be able to do everything that they might otherwise do, but they will be of enormous help. The department of social services, charged with allocating shelter, still needs physical outposts to interact with the population, but with more data it will be able to allocate beds efficiently and keep track of the resources available, among other things. A virtual military can't instill the rule of law, but it can ensure that the military and police are paid, which will assuage some fears. While governments will still be somewhat wary of entrusting their data to cloud providers, the peace of mind that backed-up institutions ensure will still be enough to justify their creation.

These institutions will offer a safety net for the population too, guaranteeing that records are preserved, employers can pay salaries, and databases of citizens both in the country and in the diaspora will be maintained. All of this will accelerate local ownership of the reconstruction process and help limit the waste and corruption that typically follow a disaster or conflict. Governments may collapse and wars can destroy physical infrastructure, but virtual institutions will survive.

Governments in exile will be capable of functioning far differently from the Polish, Belgian and French governments that were forced to operate from London during World War II. Given how well virtual institutions will function, future governments will operate remotely with a level of efficiency and reach that is unprecedented. This will be a move born of necessity, because of either a natural disaster or something more prolonged, like civil war. Imagine if Mogadishu suddenly became inoperably hostile for the beleaguered Somali government, perhaps because al-Shabaab insurgents captured the city or because clan warfare rendered the environment uninhabitable. With virtual institutions in place, government officials could relocate temporarily, inside or outside the country, and retain some semblance of control over the civil administration of the state. At a minimum, they could maintain a level of credibility with the population by arranging for salaries to be paid, coordinating with aid organizations and foreign donors, and communicating with the public in a transparent manner. Of course, virtual governance done remotely would never be anything but a last resort (surely, the distance would alter how accountable and credible the government would appear to its citizens), and certain preconditions must be in place for such a system to work, including fast, reliable and secure networks; sophisticated platforms; and a fully connected population. No state would be ready to do this today—Somalia least of all—but if countries can begin building such systems now, they will be ready when they are needed.

The potential for remote virtual governance might well affect political exiles. Whereas public figures living outside their homelands once had to rely on back channels to stay connected—the Ayatollah Khomeini famously relied on audiocassette tapes recorded in Paris and smuggled into Iran to spread his message in the 1970s—there are a range of faster, safer and more effective alternatives today. In the future, political exiles will have the ability to form powerful and competent virtual institutions, and thus entire shadow governments, that could interact with and meet the needs of the population at home.

It's not as far-fetched as it might sound. Thanks to connectivity, exiles will be far less estranged from the population than their predecessors. Acutely attuned to the trends and moods at home, they'll be able to expand their reach and influence among the population with targeted messaging on simple, popular devices and platforms. Exile leaders won't need to be concentrated in one place to form a party or movement; the differences between them that matter will be ideological, not geographic. And when these exiles have a coherent platform and vision for the country, they'll be able to transmit their plans to the population at home without ever stepping foot inside the country, quickly, securely and in so many million copies that the official government will be unable to stop the flow.

To buttress their campaign for public support, exiles will use the virtual institutions they control to win the hearts and minds of the population. Imagine a shadow government that pays and deploys an in-country security force comprising various foreign nationals to protect community strongholds, while providing e-health benefits from Paris (independent hospital administrations, coordinating free vaccination campaigns, extending virtual health-insurance plans, coordinating a network of remote doctors available for diagnostic work) and running online schools and universities from London. This government-in-exile could elect its own parliament, with campaigns and voting taking place entirely online, members drawn from several countries and sessions conducted over live-streaming video channels that can be watched by millions around the world. Even the semblance of a functioning shadow government might be enough to sufficiently sway the population at home to transfer their support from the official government to the one built and operated remotely by the exiles.

The remaining distinguishing feature of a reconstruction prototype will be close engagement with the diaspora communities.

Governments-in-exile often draw from the intellectuals in the diaspora, but the role of external communities will not only be political or financial (in the form of remittances). Connectivity means that these groups will be able to work more closely together on a much wider range of issues. The insight and depth of knowledge relevant to reconstruction possessed by members of diaspora communities is invaluable, so with greater access to communication technologies, postcrisis societies will be able to tap into those reserves of human capital in a significant way. We've already seen signs of this in some of the world's recent crises. The Somali diaspora actively used tools like Google Map Maker to identify areas affected by the 2011 drought in the Horn of Africa, using their local knowledge and connections to compile more accurate reports than outside actors could.

In the future, we will see the creation of diaspora reserve corps, with those living abroad organized by trade: doctors, police officers, construction workers, teachers and so on. States will have an incentive to organize their diaspora communities—assuming those communities are not all political exiles hostile to the state—so that they know who possesses skills that might be required in a country's time of need.

Today, several diaspora communities are far more successful than the population living back home (this includes the Iranian, Cuban and Lebanese diasporas, but also smaller groups like the Hmong and Somalis). But only portions of these communities are still connected to their native lands; many have, by choice or as a consequence of time, embraced their adoptive countries for the opportunities, security or quality of life they provide. As connectivity spreads, the gap between diaspora and home communities will shrink, as communication technologies and social media strengthen the bonds of culture, language and perspective that connect these distant groups. And those who leave their country as part of a brain drain will be leaving countries far more connected than today, even if those places are poor, autocratic or short on opportunities. Members of the diaspora, then, will be able to create a knowledge economy in exile that leverages the strong educational institutions, networks and resources of developed countries and channels them back constructively into their home countries.

Opportunism and Exploitation

In the aftermath of every major conflict or natural disaster, new actors flood the space: aid workers, journalists, U.N. officials, consultants, businessmen, speculators and tourists. Some come to offer their services, while others are hoping to exploit the crisis environment for political or economic gain. Many do both, and rather effectively so.*

Even those who don't seek financial gain have reasons beyond altruism to get involved. A postcrisis country is a great proving ground for nascent NGOs, and a platform for established nonprofit organizations to demonstrate their value to their donors. This rash of new participants—altruists and opportunists alike—can do great good, and tremendous damage. The challenge for reconstruction planners in the future will be finding ways to balance the interests and actions of all these people and groups in a productive manner.

Generally speaking, connectivity encourages and enables altruistic behavior. People have more insight and visibility into the suffering of others, and they have more opportunities to do something about it. Some scoff at the rise of "slacktivism"—slacker activism, or engaging in social activism with little or no effort—but transnational, forward-thinking organizations like Kiva, Kickstarter and Samasource represent a vision of our connected future. Kiva and Kickstarter are both crowd-funding platforms (Kiva focuses on micro-finance, while Kickstarter focuses mostly on creative pursuits), and Samasource outsources "micro-work" from corporations to people in developing countries over simple online platforms. There are other, less quantifiable ways to contribute to a distant cause than donating money, like creating supportive content or increasing public awareness, both increasingly integral parts of the process.

As more people become connected around the world, we'll see a proliferation of potential donors and activists ready to contribute to the next high-profile crisis. With real-time information about conflicts

---

* The journalist Naomi Klein famously called these actors "disaster capitalists" in her provocative book *The Shock Doctrine.* Klein argues that neo-liberal economics advocates seek to exploit a postcrisis environment to impose free-market ideals, usually to the detriment of the existing economic order. Like psychological shock therapy, this free-market fundamentalism uses the appearance of a "blank slate" to violently reshape the economic environment.

and disasters around the world increasingly accessible and available, spread evenly across different platforms in different languages, a crisis in one country can reverberate across the world instantly. Not everyone receiving the news will be spurred to action, but enough people will so that the scale of participation will rise dramatically.

Examining the aftermath of the Haiti earthquake once again will give a good indication of what the future holds. The level of destruction near the capital in Haiti, a densely populated and immensely poor country, was overwhelming: homes, hospitals and institutional buildings collapsed; transportation and communications systems were devastated; hundreds of thousands were killed and 1.5 million more made homeless.* Within hours, neighboring governments sent in emergency-services teams, and within days many countries around the world had pledged or already delivered aid.

The response from the humanitarian community was even more robust. Within days of the earthquake, the Red Cross had raised more than $5 million through an innovative "text to donate" campaign in which mobile users could text "HAITI" to a special short code (90999) to donate $10, automatically charged to their phone bill. In all, some $43 million in aid passed through mobile donation platforms, according to the Mobile Giving Foundation, which builds the technical infrastructure many NGOs used. Télécoms Sans Frontières, a humanitarian organization that specializes in emergency telecommunications, deployed on the ground in Haiti one day after the earthquake to establish call centers to allow families to reach loved ones. And just five days after the earthquake, the Thomson Reuters Foundation's AlertNet humanitarian news service set up the Emergency Information Service, the first of its kind, which allowed Haitians free SMS alert messages to help them navigate the disaster's impact.

Emergency relief efforts turned into longer-term reconstruction projects, and within months there were tens of thousands of NGOs working on the ground in Haiti. It's hard to imagine tens of thousands of aid organizations working efficiently—with clear objectives and without redundancy—in any one place, let alone a country as small,

---

* Estimates on the death toll of the Haitian earthquake vary widely. The Haitian government believes 316,000 people were killed, while a leaked memo from the U.S. government put the figure somewhere between 46,190 and 84,961.

crowded and devastated as Haiti. As the months dragged on, unsettling reports about inefficient aid distribution began to surface. Warehouses were full of unused pharmaceuticals left to expire because of poor management. Cholera outbreaks in the sprawling informal settlements threatened to wipe out many of the earthquake survivors. The delivery of funding from institutional donors, mostly governments, was delayed and difficult to keep track of; very little of the funding ever reached the Haitians themselves, having been utilized instead by any number of foreign organizations higher up on the chain. Hundreds of thousands of Haitians were still in unsanitary tent cities a year after the earthquake, because the government and its NGO partners had not yet found a way to otherwise house them. For all the coverage, the fund-raising, the coordination plans and the good intentions, Haitians were not well served in the post-earthquake environment.

People well qualified to say what transpired in Haiti have examined this fallout with great acumen—including Paul Farmer in his book *Haiti After the Earthquake*—and the consensus seems to be that this was an unfortunate confluence of factors: extensive devastation meeting bureaucratic inefficiency amid a backdrop of deeply entrenched preexisting challenges. Communication technologies could not have hoped to ameliorate all of Haiti's woes, but there are many areas where, if correctly and widely utilized, coordinated online platforms can streamline this process so that a future version of the Haitian earthquake will produce more good results and less waste in a faster recovery period. Throughout this section we will present a few of our own ideas, knowing full well that the institutional actors in reconstruction settings—the large NGOs, the foreign government donors and all the rest—may be unwilling to take these steps for fear of failure or loss of influence in the future.

As we look ahead to the next wave of disasters and conflicts that will occur in a more connected age, we can see a pattern emerging. The mixture of more potential donors and impressive online marketing will create an "NGO bubble" within each postcrisis society, and eventually that bubble will burst, ultimately leading to a greater decentralization of aid and a rash of new experiments.

Historically, what has differentiated established aid organizations is less their impact than their brand: catchy logos, poignant advertisements and prominent endorsements go much further toward attracting public donations than detailed reports about logistics, antimalarial bed nets or incremental successes. There is perhaps no better recent example of this than the now infamous *Kony 2012* video, produced by the nonprofit organization Invisible Children to generate awareness about a multi-decade-long war in northern Uganda. While the NGO's mission to end atrocities by a Ugandan militant group, the Lord's Resistance Army (LRA), was noble, many who were intimately familiar with the conflict—including many Ugandans—found the video misleading, simplistic and, ultimately, self-serving. Yet the video amassed more than 100 million views in under a week (making it the first viral video to do so), largely thanks to endorsements from prominent celebrities with millions of followers on Twitter. Early criticism of the NGO and its operations—like its 70 percent overhead in "production costs" (basically, salaries)—did little to stem the swelling movement, until it was abruptly ended by a very public and bizarre detention of one of the organization's cofounders after he exposed himself in public.

As we have already said, we will see a more level playing field for marketing in the digital era. Anyone with a registered NGO or charity (and perhaps not even that) can produce a flashy online platform with high-quality content and cool mobile apps. After all, this is the fastest and easiest way for an individual or group to make its mark. The actual substance of the organization—how robust or competent it is, how it handles finances, how good its programs might or might not be—matters less. Like certain start-up revolutionaries who value style over substance, new participants will find ways to exploit the blind spots of their supporters; in this case, these groups can take advantage of the fact that donors have little real sense of what it's like on the ground. So when a disaster strikes and NGOs pour into the space, the established ones will find themselves shoulder to shoulder with NGO start-ups, groups that have a strong online presence and starter funds but that are generally untested. Such start-ups will be more targeted in their mission than traditional aid organizations, and they'll appear equally if not more competent than their established counterparts. They'll attract attention but they'll deliver less of what is needed by those they are trying to help; some might be capable but most won't be, as they will lack the networks, the deep knowledge and the operational skills of professional organizations.

This mismatch between the start-ups' marketing and delivery will infuriate the established players. Start-ups and institutional NGOs will compete for the same resources, and the start-ups will use their digital savvy and knowledge of different online audiences to their advantage to siphon off resources from the older organizations. They'll depict the large institutional actors as lumbering, inefficient and out of touch, with high overheads, large staffs and impersonal qualities, promising instead to bring donors much closer to the recipients of aid by cutting out the middlemen. For new potential donors looking to contribute, this promise of directness will be a particularly attractive selling point since connectivity ensures that many of them will feel personally involved in the crisis already.

The concerned and altruistic young professional in Seattle with a few dollars to spare will not just "witness" every future disaster but will also be bombarded with ways to help. His inbox, Twitter feed, Facebook profile and search results will be clogged. He'll be overwhelmed but he will comb through the options and attempt to make a fast but serious judgment call based on what he sees—which group has the best-looking website, the most robust social-media presence, the highest-profile supporters. No expert, how is he to decide which organization is the right one to donate to? He'll have to rely on the trust he feels for a certain group, and in this, organizations with strong marketing skills that can pitch to him (or his profile) directly will have the edge.

There is a real risk of the traditional NGOs being crowded out by these start-up organizations. Some start-ups will be genuinely helpful, but not all will be genuine. Opportunists will take advantage of the new possibilities for direct marketing and the lower bar to entry. When those groups are eventually held to account, it will weaken donor trust (and probably generate momentum to expose more fraudulent participants). There will also be an oversupply of vanity projects from known celebrities and business leaders, whose high-wattage campaigns will only further distract attention from the real work needed to be done on the ground. In all, the result of turning "doing good" into a marketing competition means more players but less real help, as established organizations are pushed aside.

Intervention, as we've said before, requires expertise. Coordinating aid, enabling government oversight and setting realistic expectations all become harder as the field becomes more crowded. Technology can help with this. The government could keep a centralized database of all NGO actors and then register, monitor and rank each one on an online platform with the help of the public. There already are monitoring and rating systems for NGOs—Charity Navigator, One World Trust's civil-society organizations (CSO) database, and NGO Ratings—but these have mostly been NGOs themselves, even if they are helping to impose accountability, and beyond shining a spotlight on bad practices, they have no real enforcement abilities. Imagine an AAA-rating system for NGOs, where data about organizations' activities, finances and management, along with reviews from the local community and aid recipients, is used to generate a ranking that can help guide donors and their investments. The ratings would have real-world implications, including NGOs' losing eligibility for government funding if they fall below a certain score, or facing additional government scrutiny and processes. Without an integrated, transparent rating-and-monitoring system, governments and donors will come under a deluge of appeals from different aid organizations and they will have limited means to discern the legitimate and competent ones.

In the end, like all bubbles, this one will burst, as processes become delayed and institutional donors lose faith in reconstruction efforts. When the dust settles, those organizations left standing will be well-positioned NGOs with a targeted focus, strong donor loyalty and the ability to demonstrate a history of efficient and transparent operations. Some will be established aid groups and others will be new, but they will share certain characteristics that make them well suited for reconstruction work in the digital age. They will run solid programs with data-generating results, and pair their efforts on the ground with savvy digital marketing that both showcases their work and allows for responsive feedback from donors and aid recipients alike. The appearance of accountability and transparency will count for a lot.

The trend toward more direct engagement between donors and recipients on the ground will survive as well. NGOs will adopt new methods that aim to satisfy the desire to provide more intimate relationships, and in doing so they will accelerate another long-term trend visible today: the decentralization of aid distribution. By this we mean the move away from several key nodes (a few large, institutional NGOs) to networks of smaller conduits. Rather than donating to the main office of the Red Cross or Save the Children, increasingly, informed and involved donors will seek out special and specific programs that speak to them directly, or they will take their donations to smaller start-up NGOs that promise equivalent services. Smart, established NGOs will astutely reshape their function to serve more as aggregators than top-down directors, reimagining their role as one of linking donors directly to the people they fund—providing the right personal "experience," such as connecting doctors in a developed country with those in a country affected by an earthquake—while still retaining complete programmatic control. (To be sure, not all donors will seek such intimate knowledge of the organizations and individuals they support. For them, it will be easy enough to "opt out" of such engagement.)

And we cannot discount the role that individuals in countries suffering disasters or conflict will play in the newly digital aid ecosystem. Connectivity will influence how one of the biggest and most common problems that postcrisis societies face—internally displaced persons (IDPs)—will be helped. Little can be done by outsiders to prevent the conditions that lead to internal displacement within a country—war, famine, natural disaster. But mobile phones will change the future for their victims. Most dislocated people will own handsets, and if they do not (or if they have to leave them behind), relief organizations will distribute phones to them. Refugee camps will be wired with 4G hot spots that allow callers to communicate with each other easily and inexpensively, and with mobile phones, the registration of IDPs will never be easier.

Most IDPs and refugees say that among their greatest challenges is lack of information. They never know how long they'll be in one place, when food will arrive or how to get some, where they can find firewood, water and health services, and what the security threats are. With registration and specialized platforms to address these concerns, IDPs will be able to receive alerts, navigate their new environment, and receive supplies and benefits from international aid organizations on the scene. Facial-recognition software will be heavily used to find lost or missing persons. With speech-recognition technology, illiterate users will be able to speak the names of relatives and the database will report if they are in the camp system. Online platforms and mobile phones will allow refugee camps to classify and organize their members according to their skills, backgrounds and interests. In today's refugee camps, there are large numbers of people with relevant and needed skills (doctors, teachers, soccer coaches) whose participation is only leveraged in an ad hoc manner, mobilized slowly through word-of-mouth networks throughout the camps. IDPs in the future should have access to a skills-tracker app, through which they can submit their skills or search a database for what they need, leaving no skill unused or willing participant excluded.

Widespread use of mobile phones will present new opportunities for people looking to shake up the existing model of aid distribution. A few enterprising individuals with a bit of technical know-how will be able to build an open platform where potential aid recipients like themselves can list their needs and personal information, send it to the cloud and then wait for individual donors to select them and send aid directly. This is not unlike the platform that Kiva uses for micro-finance funding, except that it would be broader in scope, more personal in nature and focused on donations instead of loans. (Naturally, a platform like this would encounter a series of mechanical and legal issues that would need to be addressed before it functioned correctly.)

Now imagine if this platform partnered with a bigger organization that could promote it to a much wider audience around the world while providing some measure of verification to assuage skeptical users. In the West, a mother could take a break from watching her child's soccer game to explore a live global map (interactive and constantly updated) on her iPad, displaying who needed what and where. She would be able to independently decide whom to fund on the basis of individuals' stories or perceived need levels. Using mobile money-transfer systems already available, that mother could transfer cash or mobile credit to the recipients directly, as quickly and casually as sending a text message.

The challenge with this type of platform is that the onus of marketing falls directly on the aid recipients themselves. Life is hard enough in a refugee camp without having to worry if one's online profile is sufficiently need-worthy, and the stark competition for resources that such a platform would cause recipients is distasteful in and of itself. There is also the risk of donors who lack good judgment or familiarity with the situation on the ground disproportionately supporting people who have the best marketing campaigns (or who have gamed the system) instead of those who need it most. The consequence of going around established aid organizations is the loss of those groups' ability to discern levels of need and distribute their resources appropriately. With those controls gone, the free-for-all of direct donations would almost certainly lead to a *less* equitable division of those resources. An analysis of peer-to-peer lending through Kiva's website conducted by researchers in Singapore reported that lenders tend to discriminate in favor of attractive, lighter-skinned and less obese borrowers.

Moreover, the emergence of a platform like this assumes that the desire for a closer connection is reciprocated. Aid recipients would have to *want* to engage in such a connection, and that would strike many who have worked in development as a nonstarter. To be sure, some people in postcrisis countries (as well as developing nations) might embrace the opportunity to directly market themselves if it meant a more reliable source of funding. But the majority will not. Unlike with Kiva, whose recipients are requesting loans, these recipients would be asking for charity—publicly. Pride is a universal human quality, and often when people have little else, they value their pride all the more. It's hard to imagine that, even if such an open-funding platform were available to them, refugees, IDPs and other recipients would willingly advertise their needs to a global audience. One important function of established aid organizations is the distance they provide between recipients and their funders. So amid all of the changes we have described above—start-up NGOs, micro-targeted programs, decentralized aid—it is worth remembering the reasons certain aspects of the development-and-aid world are as they are, and why they work.

Room for Innovation

If the destruction of institutions and systems caused by upheaval has a silver lining, it's that it clears the path for new ideas. Innovation exists everywhere, even in the labored and intricate work of reconstruction, and it will be enhanced with a fast network, good leadership and plentiful devices, meaning smart phones and tablets.

We're already seeing how Internet tools are being refashioned to serve in a postcrisis environment. Ushahidi (the name means "testimony" in Swahili), an open-source crisis-mapping platform that aggregates crowd-sourced data to build a living information map, demonstrated this to great effect after the 2010 Haitian earthquake. Using a basic mapping platform, Ushahidi volunteers in the United States built a live crisis map just one hour after the earthquake struck, with a designated short code (4636) for people on the ground to text information to; it was subsequently publicized on national and local Haitian radio stations. Engineers outside Haiti added the data that was collected to an interactive online map that aggregated reports of destruction, needed emergency supplies, trapped people and violence or crime. Many of the text messages were in Creole, so Ushahidi worked with a network of thousands of Haitian-Americans to translate the information, cutting translation time to just ten minutes. Within a few weeks, they'd mapped some 2,500 reports; Carol Waters, Ushahidi-Haiti's director of communications and partnerships, said that many of those messages simply read, "I'm buried under ruble [*sic*], but I'm still alive."

Ushahidi's quick thinking and quick coding saved lives. In the future, crisis maps like these will become standard and their creation will probably be government-led. By centralizing the information with an official and trusted source, some of the problems that Ushahidi faced (like other NGOs not knowing about the platform) could be avoided. Of course, there is the risk that a government-led project would fall victim to bureaucracy or legal restrictions that would prevent it from keeping up with non-state actors like Ushahidi. But if the response were immediate, there is tremendous potential for a government-led crisis map because it could grow to encompass much more than emergency information. The map could stay active throughout the reconstruction process, and it could serve as a platform through which the government shared and received information about the various reconstruction projects and environments it managed.

For any postcrisis society, citizens could be told where known safe zones (i.e., free of mines or militia) in their neighborhoods were, where the best mobile coverage was or where the largest investments in reconstruction efforts had been made. Citizen reporting on incidents of crime, violence or corruption would keep the government informed. An integrated system of crisis information like this would not only keep the population safer, healthier and more aware, it would also cut some of the waste, corruption and redundancy that reconstruction efforts always generate. Not all postcrisis governments will be interested in such transparency, to be sure, but if the population and the international community were widely aware of the model, there might be sufficient public pressure to adopt it anyway. The delivery of foreign aid could even be made contingent on it. And no doubt there would be many willing non-state partners and volunteers ready to participate in the process.

But the first priority for a postcrisis state is, usually, managing the fragile security environment. Interactive maps can help with that, but they won't be enough. Those early moments when a conflict ends are the most delicate, because the interim government must demonstrate that it is in control *and* responsive to the people, or else it risks being chased out by the same population that installed it. In order for daily life to resume, citizens must feel safe enough to reopen businesses, rebuild homes and replant crops, so mitigating the volatility in the environment is vital for building citizen trust in the reconstruction process. Smart uses of technology can help the state reassert the rule of law in important ways.

By virtue of their functionality, mobile phones will become key conduits and valuable assets as the state works to manage the security environment. For countries with a functional military, the question of whether its members will uphold the rule of law—as opposed to defecting, committing criminal acts or seizing power for themselves—will depend less on personal motives than on their faith in the competency of the government. Put simply, for most people in uniform it will come down to whether they receive a paycheck reliably and relatively free of graft; they need to know who is in charge.

Future technology platforms will assist law enforcement in this process by equipping every police and army officer with a specialized handset device that contains several distinct (and highly secure) apps. One app will handle salaries and serve as the interface between officers and the ministry that pays them. In Afghanistan, the telecom Roshan has launched a pilot program to pay Afghan national police officers electronically through a mobile banking platform—a bold move geared toward ending the rampant corruption that cripples the country's finances. On these specialized phones, another app could require officers to report their daily activities, as they might in a logbook, storing that information in the cloud that commanders could later mine for metrics on efficiency and impact. Other apps could offer training tips or virtual mentors for newly integrated officers—as in the case of Libya, where many of the militia fighters were integrated into the newly created army—and they could provide secure online spaces for anonymous reporting of corruption or other illegal activities by other officers.

Citizen reporting over mobile platforms would strengthen the state's ability to maintain security, should the two sides choose to work together. Every citizen with a mobile device is a potential witness and investigator, more widely dispersed than any law-enforcement body and ready to document evidence of wrongdoing. In the best cases, citizens will choose to participate in these mobile vigilance activities, out of national sentiment or self-interest, and together with the state they will help build a safer and more honest society. In the worst cases, where large portions of the population distrust the government or favor the ex-combatants (like those who fought the battle against Gadhafi), those citizen-reporting channels could be used to share false information and waste police time.

Citizen engagement will be crucial beyond initial security issues, too. With the right platforms and a government inclined toward transparency, people on the ground will be able to monitor progress, report corruption, share suggestions and become an integral part of the conversations between the government, NGOs and foreign actors—all using mobile phones. We spoke with the Rwandan president Paul Kagame, who remains among the most tech-savvy leaders in Africa, and asked how mobile technology is transforming the way citizens address local challenges. "Where people have needs—economic, security and social—they will turn to their phones," he said, "because their phones are the only way to protect themselves. People who need immediate help can now get it." This, he explained, was a game-changer for populations in developing countries and particularly for people emerging from conflict or crisis. Building trust in the government is a crucial task, and by leveraging citizen participation through open platforms, this process can be much quicker and more sustainable: "In Rwanda, we have built a community policing program, where the community passes on information," Kagame said, stressing that it was made much more efficient by the use of technology.

As crowd-sourcing becomes a defining feature in the future of the rule of law—at least in the aftermath of conflict or disaster—a culture of accountability will slowly emerge. Fears of violence or looting will remain, but societies in the future will have all of their personal possessions and their historical artifacts documented online, so there won't be a question of what's missing when security returns. Citizens will be rewarded for sending in photos of thieves (even if they're police) that show their faces and their loot. The risk of retaliation would be real, but evidence suggests that despite their fear there is almost always a critical mass of people willing to take that risk. And the more people there are willing to report crime, the more the risk to the individual is reduced. Imagine if the ransacking of Iraq's celebrated Baghdad Museum in 2003 had occurred twenty years later: How long would those thieves have been able to hide their treasures (let alone try to sell them) if their theft had instantly been recorded and broadcast across the country, and other citizens were highly motivated to inform on them?

Lost artifacts damage a society's dignity and the preservation of its culture, but lost weapons constitute a far greater danger to a country's stability. Weapons and small arms routinely disappear after conflicts and find their way onto the black market (an estimated $1 billion annual business), later appearing in the hands of militias, gangs and armies in other countries. Radio frequency identification (RFID) chips could represent a solution to this challenge. RFID chips or tags contain electronically stored information and can be as small as a grain of rice. They are ever present today, in everything from our phones and passports to the products we buy. (They're even in our pets: RFID chips embedded under the skin or on an ear are used to help identify lost animals.) If major states signed treaties that required weapons manufacturers to implant unremovable RFID chips in all of their products, it would make the hunt for arms caches and the interdiction of arms shipments much easier. Given that today's RFID chips can be easily fried in a microwave, the chips of the future will need a shield that protects them against tampering. (We assume there will be a technological cat-and-mouse game between governments who want to track the weapons with RFID chips and arms traffickers who want to deal the weapons off the grid.) When weapons with RFID chips were recovered, it would be possible to trace where they'd been if the chips themselves were designed to store location data. This wouldn't stop the trafficking of arms but it would put pressure on the larger actors in the arms trade.

States that donate weapons to rebel movements often want to know what happens to those arms. With RFID chips, such investments could be tracked. The Libyan revolutionaries were an unknown quantity to almost everyone, so in the absence of any tracking capability, governments that distributed arms to them had to weigh the benefit of a successful revolution with the possible consequences of those weapons going underground. (In the beginning of 2012, some of the weaponry that Libyan militias used wound up in Mali with disgruntled Tuareg fighters. This, combined with the return of the Tuareg contingent of Gadhafi's army, led to a violent antigovernment campaign that created the conditions for a military coup.)

Electronically traceable arms distribution will have to overcome hurdles. It will cost money to design weapons that include the RFID; arms manufacturers profit from a large illicit market for their products; and states and arms dealers alike rather enjoy the anonymity of weapons distribution today. It's hard to imagine any superpower willingly sacrificing its ability to have plausible deniability regarding arms caches or covertly supplied arms for some long-term greater good. Moreover, states might claim that falsely planting another country's weapons in a conflict zone would point to their involvement and lead to even *more* conflict. But international pressure might make a difference.

Luckily, there are myriad other ways the RFID technology can be used in the short term in reconstruction efforts. RFID tags can be used to track aid deliveries and other essential supplies, to verify pharmaceuticals and other products as legitimate, and to generally limit waste or graft in large contracting projects. The World Food Program (WFP) has experimented with tracking food deliveries in Somalia, using bar codes and RFID chips to determine which suppliers are honest and deliver food to the target area. This type of tracking system—inexpensive, ubiquitous and reliable—could demonstrably help streamline the serpentine world of aid distribution by enhancing accountability and providing data that can be used to measure success and effectiveness, even in the least-connected places.

. . .

Another innovative use of mobile devices for a post-conflict government involves handling former combatants. Trading in weapons for handsets may become a key feature of any disarmament, demobilization and reintegration (DDR) program. Paul Kagame's government, while controversial in human rights and governance communities, has overseen the demilitarization of tens of thousands of former fighters through the Rwanda Demobilization and Reintegration Project. "We believe that we need to put tools in the hands of ex-combatants to transform their lives," he explained. In packages handed out to ex-combatants, "We gave them some money, but we also gave them phones so they could see what the possibilities are." Most ex-combatants coming through the still ongoing program in Rwanda also receive some form of training that will prepare them for reintegration into society. Psychological treatment is an important component as well. We've seen these programs in action, and they resemble summer camps, with classrooms, dorms and activities—fitting, since so many of the ex-combatants in Rwanda are practically children. The key is to start them off with hundreds of others who have a shared experience, and then build their confidence that there is a good life on the other side of combat.

Kagame's words indicate that we are not far away from more countries trying this. In the aftermath of every conflict, the disarmament of former combatants is a top priority. (Disarmament, sometimes referred to as demilitarization or weapons control, is the process of eliminating the military capacities of warring factions, whether they are insurgents, civil enemies or army factions left over from a previous regime.) In a typical DDR program, weapons are transferred from warring parties to peacekeeping forces over a prescribed period of time, often with some form of compensation involved. The longer the conflict, the longer it takes to complete the process. It took years of prolonged fighting between the northern and southern sides in Sudan to produce the state of South Sudan (which we had the opportunity to visit in January 2013), so the urgent need for a comprehensive DDR program was recognized immediately by the new South Sudanese government and the international community. With more than $380 million in aid from the United Nations, China, Japan, Norway and the United States, the Sudanese on both sides of the border agreed to disarm some two hundred thousand former soldiers by 2017. Two neighboring countries, Uganda and Kenya, concerned about the possible spread of combatants-turned-mercenaries and the illegal transport of arms across borders, also pledged their support in order to reinforce regional security—a critical element of the plan. However, there are few regions as unpredictable and conflict prone as the Great Lakes, so pledges must be taken with a grain of salt.

Most post-conflict environments contain armed ex-combatants who find themselves without work, purpose, status or acceptance by society. Unaddressed, these problems can lead former fighters to return to violence—as criminals, militia members or guns for hire—especially if they still have their weapons. As governments seek to create incentives for ex-combatants to turn in their AK-47s, they will find that the prospect of a smart phone might be more than enough to get started. Former fighters need compensation, status and a next step. If they are made to understand that a smart phone represents not just a chance to communicate but also a way to receive benefits and payment, the phone becomes an investment that is worth trading a weapon for.

Each society will offer slightly different packages in this initiative, depending on the culture and the level of technological sophistication, but the essentials of the process have a universal appeal: free top-of-the-line devices, cheap text and voice plans, credit to purchase apps, and data subsidization that allows people to use the Internet and e-mail inexpensively. These smart phones would be of a better quality than much of the population's and cheaper to use, as well. They could be front-loaded with appealing vocational applications that would provide some momentum for upwardly mobile ex-combatants, like English-language instruction or even basic literacy education. A former child soldier in a South Sudanese refugee camp, who had been forced to leave his family at a young age, could have access to a device that connected him not only to local relatives, but also to potential mentors from the Sudanese diaspora abroad, perhaps young men who had successfully sought asylum in the United States and built wholly new lives for themselves.

Donor nations would likely pay for a program like this in its initial stages, then transfer the cost and control to the state in question. That would allow the government to maintain some leverage over the ex-combatants in its society. There could be software preloaded on the phone that allowed the state to track ex-combatants or monitor their browsing history for some period of time; ex-combatants would risk losing the data plan or the phone if they didn't follow the rules of the program. A state would be able to institute a three-strike policy tied to the geo-location data on these phones: The first time an ex-combatant failed to check in with his equivalent of a probation officer at a pre-scribed time, he would receive a short video warning; the second fail-ure would result in the data plan being suspended for some length of time, and the third failure would lead to the cancellation of the data plan and the repossession of the device.

Of course, enforcement would be a challenge, but the state would at least have more leverage than it would from a one-time cash payment. And there are ways to make this program desirable beyond useful apps and status-symbol phones. Ex-combatants will likely rely on pensions or benefits to provide for their families, so integrating those payments into a mobile money system is a smart way to keep the former fighters on the right path.

In order for this arms-for-phones project to work, however, it would need to be tied to a comprehensive and successful program—mobile phones alone would not get thousands of former fighters reintegrated in any sustainable way. As part of the reintegration and accountability programs, some ex-combatants would receive cash or special features for their device in exchange for photographs of arms caches or mass graves. Ex-combatants would have to feel fairly treated and adequately compensated to surrender both their guns and their sense of author-ity; programs that included counseling and classes in job skills would be important for helping these individuals transition into civilian life.

In Colombia, a largely successful DDR program to reintegrate for-mer guerrilla fighters into society involved a wide network of support centers for ex-combatants, offering them educational, legal, psycho-social and health services. Unlike many other DDR programs, which are run far away from city centers, the government of Colombia made the bold move of placing many of the reintegration houses in the mid-dle of the city. The government identified a need early on to build confidence in the program, both on the ex-combatant side and within society. Set up much like homes for runaway teens, these houses eventually became part of the community, with neighbors and other locals getting involved. The government used ex-combatants as spokespersons for why Colombians should not turn to violence. They spoke at universities, addressed former members of the Revolutionary Armed Forces of Colombia (FARC)—a forty-eight-year-old Colombian terrorist organization—and conducted community roundtables.

It's unclear whether communication technologies will help or hinder the reconciliation process for noncombatants. On one hand, the ubiquity of devices during a conflict will help empower citizens to capture evidence they can use to seek justice in the post-conflict environment. On the other hand, with so much violence and suffering caught on digital tape (stored in perpetuity, and shared widely), it's possible that the social or ethnic divisions that engendered the conflict will solidify when the volume of data is brought to light. The healing process for societies torn apart by civil or ethnic conflict is painful enough, and it requires a certain collective memory loss. With much more evidence, there will be much more to forgive.

In the future, technology will be used to document and record the implementation of various transitional justice processes, including reparations, vetting (like de-Baathification efforts), truth-and-reconciliation commissions, and even trials, making each of them more accessible and transparent. There are good and bad aspects to this shift. The televised trial of Saddam Hussein was cathartic for many Iraqis, but it also gave the late dictator and his supporters a stage on which to perform. Then again, as Nigel Snoad, a former senior U.N. aid worker now at Google, predicted, "Human-rights and justice groups can build a system for people to create memorials and to tell the story of those killed and who disappeared in the conflict." Using these testimonials and memorials, he said, groups could "bring together stories from both sides, and despite conflicting accounts and occasional online flame wars (character bashing over the Internet through discussion lists and comments), create a space for apologies, truth-telling and an emerging reconciliation."

The slow, painful mechanics of reconciliation will not be eliminated by Internet technology, nor should they be. Public admissions of guilt, sentencing and punishment, and gestures of forgiveness, are all cathartic for a society recovering from conflict. Today's models for criminal prosecution at the international level—for crimes against humanity—

are slow, bureaucratic and prone to corruption. Dozens of criminals sit in the International Criminal Court (ICC)—more casually referred to as The Hague—for many months before their trials even start. In today's post-conflict environments, local court systems and indigenous local bodies are frequently preferred over the international institutions that lag behind.

The spread of technology is likely to exacerbate this trend. The sheer volume of digital evidence of crimes and violence will raise expectations that justice must be done, yet the glacial pace displayed by international judicial bodies like the ICC will limit how quickly such bodies adapt to these changes. For example, the ICC is unlikely to ever accept unverified videos captured on a mobile phone as evidence in its highly procedural trials (although organizations like Witness are trying to challenge this), but local judicial systems, with fewer legal constraints and a more flexible attitude, might be more open to developments in digital watermarking that will allow firsthand videos to be effectively authenticated. People will increasingly show their preference for these judicial avenues.

A local setting means that adjudicators, whether they are formal judges, tribal chiefs or community leaders, must have an intimate and expansive knowledge of the society—internal dynamics, main actors, major villains and all the nuances that international or distant bodies struggle to understand. When presented with digital evidence, the need for verification is lower, because the people and places are already familiar. In a postcrisis setting, there is also a distinct pressure from the community to mete out justice quickly. Whether these courts would be more or less fair than their international counterparts is a matter of debate, but they'll surely move faster.

This trend could be manifested in future truth-and-reconciliation committees, or in temporary judicial structures after a major conflict. After the Rwandan genocide, the country's new government rejected the South African truth-and-reconciliation model, arguing that reconciliation would take place only when the guilty were punished. But the formal judicial system took too long to process alleged *genocidaires;* more than a hundred thousand Rwandans sat in jail for several years waiting for their time in court. So a new system of local courts was built, taking inspiration from a grassroots, community-based conflict-resolution process known as *"gacaca."* Under *gacaca* tribunals, the accused were confronted by the community and offered a commuted sentence if they confessed their crimes, shed light on what happened or identified the remains of those they killed. Despite being based in village justice, the *gacaca* tribunal system was a complex structure, involving different phases for judgment. The first phase was referred to as the cell level; in it the accused were brought before a tribunal of people in the community where the crime was committed. This tribunal determined the severity of the crime—whether the accused should be tried at the sector, district or province level, all three of which deal with appeals. The *gacaca* system was far from perfect. It came with the full panoply of traditional cultural prejudices, including the exclusion of women as judges and a failure to prosecute crimes committed against women with the same ferocity as those against men. These caveats aside, justice was fast, and the participating community generally felt satisfied with the process. Subsequent postcrisis governments elsewhere in the world have looked at adopting this model given how effective it was at advancing numerous reconciliation goals.

Whether citizens in the future choose to take their digital evidence to The Hague or to local judicial bodies, they will certainly have more opportunities to participate in the transitional justice-and-reconciliation process. They can instantly upload documents, photos and other evidence from a conflict or a former repressive regime to an international cloud-based data bank that will categorize and add the information to the relevant open files, to be used later by courts, journalists and others. Participatory memorials and inclusive feedback loops that allow populations to express their grievances in an organized manner—perhaps communities will use algorithmic argument mapping to aggregate the most prescriptive feedback—will help retain the confidence of groups that, once a conflict is over, might begin to feel neglected. Citizens will be able to watch the justice process unfold in real time, with live-streaming trials of major figures halfway across the world available on their phones, and a wealth of information about each stage of the process at their fingertips. Documenting the crimes (both physical and virtual) of a fallen regime serves a broader purpose beyond prosecution: Once every dirty secret of the former state is published online, no future government will be able to do quite the same things. Political observers always worry about a post-conflict state's slide back into autocracy and watch keenly for signs of such a return;

the full exposure of the former regime's wrongdoings—how exactly it brutalized dissidents, how it spied on citizens' online activities, how it hid money out of the country—will help forestall such possibilities.

Among all of the topics we've covered, the future of reconstruction is perhaps where the greatest share of optimism belongs. Little can be more devastating to a country and a population than natural disaster or war, or both, and yet we see a clear trend of postcrisis transitions occurring in shorter time periods with more satisfactory results. Unlike many avenues in geopolitics, the world does learn from each reconstruction example what works, what doesn't and what can be improved upon. Clever applications of communications technology and widespread connectivity will accelerate rebuilding, inform and empower the people, and help forge a better, stronger and more resilient society. All it takes is a bit of creativity, plenty of bandwidth and the will to innovate.

Conclusion

As we look into the future—its promises and its challenges—we are facing a brave new world, the most fast-paced and exciting period in human history. We'll experience more change at a quicker rate than any previous generation, and this change, driven in part by the devices in our own hands, will be more personal and participatory than we can even imagine.

In 1999, the futurist Ray Kurzweil proposed a new "Law of Accelerating Returns" in his seminal book *The Age of Spiritual Machines: When Computers Exceed Human Intelligence.* "Technology," he wrote, "is the continuation of evolution by other means, and is itself an evolutionary process." Evolution builds on its own increasing order, leading to exponential growth and accelerated returns over time. Computation, the backbone of every technology we see today, behaves in much the same way. Even with its eventual inevitable limitations, Moore's Law promises us infinitesimally small processors in just a matter of years. Every two days we create as much digital content as we did from the dawn of civilization until 2003—that's about five exabytes of information, with only two billion people out of a possible seven billion online. How many new ideas, new perspectives and new creations will truly global technological inclusion produce, and how much more quickly will their impact be felt? The arrival of more people in the virtual world is good for them, and it's good for us. The collective benefit of sharing human knowledge and creativity grows at an exponential rate.

In the future, information technology will be everywhere, like electricity. It will be a given, so fully a part of our lives that we will struggle to describe life before it to our children. As connectivity ushers billions more people into the technological fold, we know that technology will soon be intertwined with every challenge in the world. States, citizens and companies will make it part of every solution.

Attempts to contain the spread of connectivity or curtail people's access will always fail over a long enough period of time—information, like water, will always find a way through. States, citizens, companies, NGOs, consultants, terrorists, engineers, politicians and hackers will all try to adapt to this change and manage its aftereffects, but none will be able to control it.

We believe the vast majority of the world will be net beneficiaries of connectivity, experiencing greater efficiency and opportunities, and an improved quality of life. But despite these almost universal benefits, the connected experience will not be uniform. A digital caste system will endure well into the future, and people's experience will be greatly determined by where they fall in this structure. The tiny minority at the top will be largely insulated from the less enjoyable consequences of technology by their wealth, access or location. The world's middle class will drive much of the change, as they'll be the inventors, the leaders in diaspora communities and the owners of small and medium-sized enterprises. These are the first two billion who are already connected.

The next five billion people to join that club will experience far more change, simply because of where they live and how numerous they are. They'll receive the greatest benefits from connectivity but also face the worst drawbacks of the digital age. It is this population that will drive the revolutions and challenge the police states, and they'll also be the people tracked by their governments, harassed by online hate mobs and disoriented by marketing wars. Many of the challenges in their world will endure even as technology spreads.

So, what do we think we know about our future world?

First, it's clear that technology alone is no panacea for the world's ills, yet smart uses of technology can make a world of difference. In the future, computers and humans will increasingly split duties according to what each does well. We will use human intelligence for judgment, intuition, nuance and uniquely human interactions; we will use computing power for infinite memory, infinitely fast processing and actions limited by human biology. We'll use computers to run predictive correlations from huge volumes of data to track and catch terrorists, but how they are interrogated and handled thereafter will remain the purview of humans and their laws. Robots in combat will prevent deaths through greater precision and situational awareness, but human judgment will determine the context in which they are used and what actions they can take.

Second, the virtual world will not overtake or overhaul the existing world order, but it will complicate almost every behavior. People and states will prefer the worlds where they have more control—virtual for people, physical for states—and this tension will exist as long as the Internet does. Crowds of virtually courageous people might be sufficient to start a revolution, but the state can still use brutal tactics in crackdowns on the street. Minority groups might pursue virtual statehood and cement their solidarity in the process, but if the venture goes badly, participants and their cause could end up worse off in both the physical and the virtual world as a result.

Third, states will have to practice two foreign policies and two domestic policies—one for the virtual world and one for the physical world—and these policies may appear contradictory. States will launch cyber attacks against countries they wouldn't dream of targeting militarily. They'll allow for the venting of dissent online, but viciously patrol the town square looking for vocal dissidents to crack down on. States will support emergency telecommunications interventions without even considering putting boots (or bots) on the ground.

Finally, with the spread of connectivity and mobile phones around the world, citizens will have more power than at any other time in history, but it will come with costs, particularly to both privacy and security. The technology we talk about collects and stores much personal information—past, present and future locations as well as the information you consume—all stored for a time for the systems to work. Such information has never been available before, and there is always the potential that it could be used against you. Nations will legislate much of this and their policies will differ, not just from democracy to autocracy, but even within countries that have similar political systems. The risk that this information may be released is increasing, and while the technology to protect it is available, human error, nefarious activity and the passage of time means that it will become only more difficult to keep information private. The companies responsible for storing this data have a responsibility to ensure its security, and that will not change. While the protection of individual privacy is also their responsibility, it is one that they share with the users.

We need to fight for our privacy or we will lose it, particularly in moments of national crisis, when security hawks will insist that with each terrible crime, governments are entitled to access more private, or formerly private, information. Governments have to decide where the new privacy line is, and stick to it. Facial recognition, for example, will keep people safe and ensure that they count in everything from a census to a vote, by making it easier to catch and capture illicit actors, discouraging would-be criminals and promoting public safety. But it can also empower governments to exercise greater surveillance of their people.

And what of the prospects for keeping secrets in the future, something equally important for the proper functioning of people and institutions? New abilities to encrypt secrets and spread pieces of information among people will lead to some unusual new problems. Separate groups—ranging from criminals to dissidents—will soon be able to take a secret (perhaps a set of codes or classified documents), encrypt it and then divide up the secret by allocating one part of the encryption key to each group member. A group could then consent to a mutually assured publication pact—that is, under certain circumstances, everyone combines his partial key to release the data. Such an agreement could be used to discipline governments or terrorize individuals. And if groups like al-Qaeda get their hands on sensitive encrypted data—such as the names and locations of undercover CIA agents—they could distribute copies to their affiliates with a common key and threaten to release the information if any one of their groups is attacked.

What emerges in the future, and what we've tried to articulate, is a tale of two civilizations: One is physical and has developed over thousands of years, and the other is virtual and is still very much in formation. These civilizations will coexist in a more or less peaceable manner, with each restraining the negative aspects of the other. The virtual world will enable escape from the repression of state control, offering citizens new opportunities to organize and revolt; other citizens will simply connect, learn and play. The physical world will impose rules and laws that help contain the anarchy of virtual space and that protect people from terrorist hackers, misinformation and even from the digital records of their own youthful misbehavior. The permanence of evidence will make it harder for the perpetrators of crimes to minimize or deny their actions, forcing accountability into the physical world in a way never before seen.

The virtual and physical civilizations will affect and shape each other; the balance they strike will come to define our world. In our view, the multidimensional result, though not perfect, will be more egalitarian, more transparent and more interesting than we can even imagine. As in a social contract, users will voluntarily relinquish things they value in the physical world—privacy, security, personal data—in order to gain the benefits that come with being connected to the virtual world. In turn, should they feel that these benefits are being withheld, they'll use the tools at their disposal to demand accountability and drive change in the physical world.

The case for optimism lies not in sci-fi gadgets or holograms but in the check that technology and connectivity bring against the abuses, suffering and destruction in our world. When exposure meets opportunity, the possibilities are endless. The best thing anyone can do to improve the quality of life around the world is to drive connectivity and technological opportunity. When given the access, the people will do the rest. They already know what they need and what they want to build, and they'll find ways to innovate with even the meagerest set of tools. Anyone passionate about economic prosperity, human rights, social justice, education or self-determination should consider how connectivity can help us reach these goals and even move beyond them. We cannot eliminate inequality or abuse of power, but through technological inclusion we can help transfer power into the hands of individual people and trust that they will take it from there. It won't be easy, but it will be worth it.

ACKNOWLEDGMENTS

This book is the product of nearly three years of collaboration, but it would not have been possible without the incredibly generous commitments made by close friends, family and colleagues.

First and foremost, we owe a huge debt of gratitude to Sophie Schmidt, who served as our internal editor on the book for ten months and was a critical partner in its writing. Sophie's gifted mind, strategic insights and analytical heft helped make the ideas come alive. Her grasp of both the political and the technological worlds uniquely positioned her to help ensure that the book had the right rigor and appropriate balance between tech and foreign policy on the one hand, and present-day analysis and futuristic speculation on the other. Sophie also joined us as part of a traveling trio to a number of the global hot spots that we write about.

We also owe a big thanks to the Council on Foreign Relations (CFR), who first suggested that we write a piece together for *Foreign Affairs* in the summer of 2010. That article inspired conversations that led to this book. Special thanks to Richard Haass and the other CFR executives.

We are grateful to our friend Scott Malcomson, who in the early days of the manuscript proved to be an indispensable partner and editorial advisor. Before engaging Scott, we were both admirers of his work as a journalist, foreign-policy thinker and author. His deep generalist knowledge, expertise on the international system and appreciation for the disruptive nature of technology made him the perfect advisor and editor during the critical early drafting stages. What we are most grateful for, however, is the friendship we built with such a wonderful and brilliant person throughout this process.

A special thanks to our first readers of the manuscript: Robert Zoellick, Anne-Marie Slaughter, Michiko Kakutani, Alec Ross and Ian Bremmer. Each of them took time out of his or her very busy schedule to give in-depth feedback and professional perspectives.

We had several research associates, without whom this book would not have been possible. Special thanks to Kate Krontiris, who helped ensure that our boldest claims were rooted in proper quantitative data. We also want to thank Andrew Lim, who was tireless in the research he did, which proved to be relevant to every chapter. Andrew's ability to conduct thorough research almost overnight impressed both of us. We also want to thank Thalia Beaty, who joined us toward the end and was hugely helpful on some of the final research.

Personal interviews proved invaluable, and we want to thank in particular former secretary of state Henry Kissinger; President Paul Kagame of Rwanda; Prime Minister Mohd Najib Abdul Razak of Malaysia; Mexico's former president Felipe Calderon; the Saudi prince Al-Waleed bin Talal; Ashfaq Parvez Kayani, Chief of Army Staff of the Pakistan Army; Shaukat Aziz, former prime minister of Pakistan; WikiLeaks' cofounder Julian Assange; Mongolia's former prime minister Sukhbaatar Batbold; the Mexican businessman Carlos Slim Helú; Prime Minister Hamadi Jebali of Tunisia; the former DARPA administrator turned Googler Regina Dugan; Android's senior vice-president Andy Rubin; Microsoft's chief research officer, Craig Mundie; Vodafone's CEO, Vittorio Colao; the Brookings senior fellow Peter Singer; former Mossad chief Meir Dagan; Taj Hotels' CIO, Prakash Shukla; and the former Mexican secretary of the economy Bruno Ferrari.

We had a number of friends, colleagues and family who allowed us to impose on them at various stages of the writing process. We'd like to thank Pete Blaustein, a rising star in the field of economics, whose insights proved essential to several chapters of this book; Jeffrey McLean, who offered invaluable strategic insights into the future of combat and conflict; Trevor Thompson, who helped us better understand the future battlefield; and Nicolas Berggruen, who was one of our early motivators in the development of this book and who read some of our earliest drafts.

Knopf is an amazing publisher, and it is easy to see where its reputation comes from. Its leader, Sonny Mehta, encouraged us to be bold, think big and write something that would look forward. Jonathan Segal more than lived up to his reputation, helping us take the manuscript in directions that made it much stronger. His creativity and vision as an editor were critical to making the book possible. Our thanks to Paul Bogaards, Maria Massey and Erinn Hartman, consumate professionals all.

Our agent, Mel Parker, ensured that we found a publisher who shared our vision in tackling these difficult issues. We would also like to thank the many people at Google who offered their important insights at various stages in the writing process. Google's cofounders Larry Page (also CEO) and Sergey Brin are a constant source of inspiration for both of us. Justin Kosslyn, a product manager at Google Ideas and a product visionary, helped us shape several of our future predictions. Justin is undoubtedly going to be someone to watch in the future. Lucas Dixon, an associate on the Google Ideas team and a brilliant engineer, helped us work through some of the more technical aspects of the book. We also benefited from conversations with many current and former Googlers: CJ Adams, Larry Alder, Nikesh Arora, Jieun Baek, Brendan Ballou, Andy Berndt, Eric Brewer, Shona Brown, Scott Carpenter, Christine Chen, DJ Collins, Yasmin Dolatabadi, Marc Ellenbogen, Eric Gross, Jill Hazelbaker, Shane Huntley, Minnie Ingersoll, Amy Lambert, Ann Lavin, Erez Levin, Damian Menscher, Misty Muscatel, David Pressoto, Scott Rubin, Nigel Snoad, Alfred Spector, Matthew Stepka, Astro Teller, Sebastian Thrun, Lorraine Twohill, Rachel Whetstone, Mike Wiacek, Susan Wojcicki and Emily Wodd.

There are a number of people at Google who helped orchestrate many of the logistics and trips that helped make this book possible: Jennifer Barths, Kimberly Birdsall, Gavin Bishop, Kimberly Cooper, Daniela Crocco, Dominique Cunningham, Danielle "Mr. D" Feher, Ann Hiatt, Dan Keyserling, Marty Lev, Pam Shore, Manuel Temez and Brian Thompson.

Our gratitude to all our friends and colleagues whose ideas and thoughts we've benefited from: Elliott Abrams, Ruzwana Bashir, Michael Bloomberg, Richard Branson, Chris Brose, Jordan Brown, James Bryer, Mike Cline, Steve Coll, Peter Diamandis, Larry Diamond, Jack Dorsey, Mohamed El-Erian, James Fallows, Summer Felix, Richard Fontaine, Dov Fox, Tom Freston, Malcolm Gladwell, James Glassman, Jack Goldsmith, David Gordon, Sheena Greitens,

Craig Hatkoff, Michael Hayden, Chris Hughes, Walter Isaacson, Dean Kamen, David Kennedy, Erik Kerr, Parag Khanna, Joseph Konzelmann, Stephen Krasner, Ray Kurzweil, Eric Lander, Jason Liebman, Claudia Mendoza, Evgeny Morozov, Dambisa Moyo, Elon Musk, Meghan O'Sullivan, Farah Pandith, Barry Pavel, Steven Pinker, Joe Polish, Alex Pollen, Jason Rakowski, Lisa Randall, Condoleezza Rice, Jane Rosenthal, Nouriel Roubini, Kori Schake, Vance Serchuk, Michael Spence, Stephen Stedman, Dan Twining, Decker Walker, Matthew Waxman, Tim Wu, Jillian York, Juan Zarate, Jonathan Zittrain and Ethan Zuckerman.

We also want to thank the guys from Peak Performance, particularly Joe Dowdell and Jose and Emilio Gomez, for keeping us healthy during the final stages of writing.

And to our families: From Jared, a very special thank-you to Rebecca Cohen, who during our writing process went from being a long-distance girlfriend to a wife. Throughout, she has been an intellectual partner, and served as one of our most helpful advisors. Her expertise and knowledge of the legal system brought up a number of provocative questions that ended up becoming defining features of several chapters. Also a special thanks to Dee and Donald Cohen, Emily and Jeff Nestler, Annette and Paul Shapiro, Audrey Bear, and Aaron and Rachel Zubaty for being such a supportive family. There is also a special debt of gratitude owed to Alan Mirken, who is a veteran of the publishing industry and in addition to being a great uncle (pun intended), is always insightful in his advice and guidance.

From Eric, a lifetime of thank-yous to Wendy Schmidt, who brought a sense of humanity and purpose to a dry technology executive. She bridges the human and technological worlds flawlessly.

—E.S., J.C., January 2013

NOTES

Introduction

3   The Internet is among the few things: This quote is adapted from part of Eric Schmidt's speech at the April 1997 JavaOne Conference in San Francisco. The original quote is "The Internet is the first thing that humanity has built that humanity doesn't understand, the largest experiment in anarchy that we have ever had." We have adapted the quote to our current view, which is that it is not the first thing, but instead "among the few," with others including nuclear weapons, steam power, and electricity.

4   it is the first that will make it possible: The printing press, the landline, the radio, the television, and the fax machine all represent technological revolutions, but all required intermediaries.

4   350 million: See figures for year 2000 in "Estimated Internet Users (World) and Percentage Growth," ITU World Telecommunication Indicators (2001), referred to by Claudia Sarrocco and Dr. Tim Kelly, *Improving IP Connectivity in the Least Developed Countries,* International Telecommunication Union (ITU), Strategy and Policy Unit, 9, accessed October 23, 2012, http://www.itu.int/osg/spu/ni/ipdc/study /Improving%20IP%20Connectivity%20in%20the%20Least%20Developed%20 Countries1.pdf.

4   more than 2 billion: See figures for year 2010 in "Global Numbers of Individuals Using the Internet, Total and Per 100 Inhabitants, 2001–2011," International Telecommunication Union (ITU), ICT Data and Statistics (IDS), accessed October 8, 2012, http://www.itu.int/ITU-D/ict/statistics/.

4   from 750 million to well over 5 billion: See sums for years 2000 and 2010 in "Mobile-Cellular Telephone Subscriptions," International Telecommunication Union (ITU), ICT Data and Statistics (IDS), accessed October 8, 2012, http://www.itu.int/ITU -D/ict/statistics/.

4   projected eight billion: See total for both sexes' population in "World Midyear Population by Age and Sex for 2025," U.S. Census Bureau, International Data Base, accessed October 8, 2012, http://www.census.gov/population/international/data /idb/worldpop.php.

6   many old institutions . . . reallocate the concentration of power: This concept was something we had discussed for a while, but it wasn't until a conversation with our good friend Alec Ross that we were able to capture it in this way. He deserves shared credit for this concept. See Alec Ross, "How Connective Tech Boosts Political

Change," CNN, June, 20, 2012, http://www.cnn.com/2012/06/20/opinion/opinion
-alec-ross-tech-politics/index.html.

8   banned the use of mobile phones: "Better than Freedom? Why Iraqis Cherish Their
       Mobile Phones," *Economist,* November 12, 2009, http://www.economist.com/node
       /14870118.

8   unreliable access to food, water and electricity: "Iraq: Key Facts and Figures," BBC,
       September, 7, 2010, http://www.bbc.co.uk/news/world-middle-east-11095920.

8   garbage hadn't been collected in *years*: Zaineb Naji and Dawood Salman, "Bagh-
       dad's Trash Piles Up," Environmental News Service, July 6, 2010, http://www.ens
       -newswire.com/ens/jul2010/2010-07-06-01.html.

CHAPTER 1   Our Future Selves

13   five billion more people: *The World in 2011: ICT Facts and Figures,* International
       Telecommunication Union (ITU), accessed October 10, 2012, http://www.itu.int
       /ITUD/ict/facts/2011/material/ICTFactsFigures2011.pdf. The above source shows
       that as of 2011 35 percent of the world's population is online. We factored in popula-
       tion increase projections to estimate five billion set to join the virtual world.

14   Consider the impact of basic mobile phones: This fisherwomen thought experiment
       came out of a conversation with Rebecca Cohen, and while we put it in the context
       of the Congo, the example belongs to her.

14   650 million mobile-phone users in Africa: "Africa's Mobile Phone Industry 'Boom-
       ing,'" BBC, November 9, 2011, http://www.bbc.co.uk/news/world-africa-15659983.

14   close to 3 billion across Asia: See mobile cellular subscriptions, Asia & Pacific, year
       2011, in "Key ICT Indicators for the ITU/BDT Regions (Totals and Penetration
       Rates)," International Telecommunication Union (ITU), ICT Data and Statistics
       (IDS), updated November 16, 2011, http://www.itu.int/ITU-D/ict/statistics/at
       _glance/KeyTelecom.html.

14   The majority of these people are using basic-feature phones: Ibid. Compare mobile
       cellular subscriptions to active mobile broadband subscriptions for 2011.

14   life expectancy is less than sixty years, or even fifty: "Country Comparison: Life
       Expectancy at Birth," CIA, World Fact Book, accessed October 11, 2012, https://www
       .cia.gov/library/publications/the-world-factbook/rankorder/2102rank.html#top.

15   This will even be true: One of the authors spent the summer of 2001 in this remote
       village, without electricity, running water, or a single cell phone or landline. During
       a return trip in the fall of 2010, many of the Maasai women had crafted beautiful
       beaded pouches to store their cell phones in.

15   China's expansive *"shanzhai"* network: Nicholas Schmidle, "Inside the Knockoff-
       Tennis-Shoe Factory," *New York Times Magazine,* August 19, 2010, Global edition,
       http://www.nytimes.com/2010/08/22/magazine/22fake-t.html?pagewanted=all.

15   machines can actually "print" physical objects: "The Printed World: Three-
       Dimensional Printing from Digital Designs Will Transform Manufacturing and
       Allow More People to Start Making Things," *Economist,* February 10, 2011, http://
       www.economist.com/node/18114221.

16   a full-sized replica motorcycle: Patrick Collinson, "Hi-Tech Shares Take US for
       a Walk on the High Side," *Guardian* (Manchester), March 16, 2012, http://www
       .guardian.co.uk/money/2012/mar/16/hi-tech-shares-us.

18   "social robots" that can recognize human gestures: Sarah Constantin, "Gesture

Recognition, Mind-Reading Machines, and Social Robotics," *H+ Magazine,* February 8, 2011, http://hplusmagazine.com/2011/02/08/gesture-recognition-mind-reading-machines-and-social-robotics/.

18   In 2012, a team at a robotics laboratory in Japan: Helen Thomson, "Robot Avatar Body Controlled by Thought Alone," *New Scientist,* July 2012, 19–20.

20   Consider the twenty-four-year-old Kenyan inventor Anthony Mutua: "Shoe Technology to Charge Cell Phones," *Daily Nation,* May 2012, http://www.nation.co.ke/News/Shoe+technology+to+charge+cell+phones++/-/1056/1401998/-/view/printVersion/-/sur34lz/-/index.html.

20   placed the chip in the sole of a tennis shoe: Ibid.

20   Mutua's chip is now set to go into mass production: Ibid.

21   Khan Academy: In the spirit of full disclosure: Eric Schmidt is on the board of Khan Academy.

21   replacing lectures with videos watched at home: Clive Thompson, "How Khan Academy Is Changing the Rules of Education," *Wired Magazine,* August 2011, posted online July 15, 2011, http://www.wired.com/magazine/2011/07/ff_khan/.

22   In 2012, the MIT Media Lab tested: Nicholas Negroponte, "EmTech Preview: Another Way to Think About Learning," *Technology Review,* September 13, 2012, http://www.technologyreview.com/view/429206/emtech-preview-another-way-to-think-about/.

22   distributing preloaded tablets to primary-age kids: David Talbot, "Given Tablets but No Teachers, Ethiopian Children Teach Themselves," *Technology Review,* October 29, 2012, http://www.technologyreview.com/news/506466/given-tablets-but-no-teachers-ethiopian-children-teach-themselves/.

22   one of the lowest rates of literacy in the world: "Field Listing: Literacy," CIA, World Fact Book, accessed October 11, 2012, https://www.cia.gov/library/publications/the-world-factbook/fields/2103.html#af.

25   in 2012, Nevada became the first state to issue licenses to driverless cars: Chris Gaylord, "Ready for a Self-Driving Car? Check Your Driveway," *Christian Science Monitor,* June 25, 2012, http://www.csmonitor.com/Innovation/Tech/2012/0625/Ready-for-a-self-driving-car-Check-your-driveway.

25   California also affirmed their legality: James Temple, "California Affirms Legality of Driverless Cars," *The Tech Chronicles* (blog), *San Francisco Chronicle,* September 25, 2012, http://blog.sfgate.com/techchron/2012/09/25/california-legalizes-driverless-cars/; Florida has passed a similar law. See Joann Muller, "With Driverless Cars, Once Again It Is California Leading the Way," *Forbes,* September 26, 2012, http://www.forbes.com/sites/joannmuller/2012/09/26/with-driverless-cars-once-again-it-is-california-leading-the-way/.

26   Food and Drug Administration (FDA) approved the first electronic pill in 2012: Erin Kim, "'Digital Pill' with Chip Inside Gets FDA Green Light," *CNN Money,* August 3, 2012, http://money.cnn.com/2012/08/03/technology/startups/ingestible-sensor-proteus/index.htm; Peter Murray, "No More Skipping Your Medicine—FDA Approves First Digital Pill," *Forbes,* August 9, 2012, http://www.forbes.com/sites/singularity/2012/08/09/no-more-skipping-your-medicine-fda-approves-first-digital-pill/.

26   pill carries a tiny sensor one square millimeter in size: Ibid.

26   stomach acid activates the circuit: Daniel Cressey, "Say Hello to Intelligent Pills: Digital System Tracks Patients from the Inside Out," *Nature,* January 17, 2012, http://www.nature.com/news/say-hello-to-intelligent-pills-1.9823; Randi Martin,

"FDA Approves 'Intelligent' Pill That Reports Back to Doctors," *WTOP*, August 2, 2012, http://www.wtop.com/267/2974694/FDA-approves-intelligent-pill -that-reports-back-to-doctors.

26   The patch can collect information: Cressey, "Say Hello to Intelligent Pills," *Nature*, January 17, 2012, and Martin, "FDA Approves 'Intelligent' Pill," *WTOP*, August 2, 2012.

26   track what a person eats: Randi Martin, "FDA Approves 'Intelligent' Pill That Reports Back to Doctors," *WTOP*, August 2, 2012.

26   Tissue engineers will be able to grow new organs: Henry Fountain, "One Day, Growing Spare Parts Inside the Body," *New York Times*, September 17, 2012, http://www .nytimes.com/2012/09/18/health/research/using-the-body-to-incubate-replacement -organs.html?pagewanted=all; Henry Fountain, "A First: Organs Tailor-Made with Body's Own Cells," *New York Times*, September 15, 2012, http://www.nytimes .com/2012/09/16/health/research/scientists-make-progress-in-tailor-made-organs .html?pagewanted=all; Henry Fountain, "Synthetic Windpipe Is Used to Replace Cancerous One," *New York Times*, January 12, 2012, http://www.nytimes.com /2012/01/13/health/research/surgeons-transplant-synthetic-trachea-in-baltimore -man.html.

27   doctors and disease specialists will have more information: Gina Kolata, "Infant DNA Tests Speed Diagnosis of Rare Diseases," *New York Times*, October 3, 2012, http://www.nytimes.com/2012/10/04/health/new-test-of-babies-dna-speeds -diagnosis.html?_r=1; Gina Kolata, "Genome Detectives Solve a Hospital's Deadly Outbreak," *New York Times*, August 22, 2012, http://www.nytimes.com/2012/08 /23/health/genome-detectives-solve-mystery-of-hospitals-k-pneumoniae-outbreak .html; Gina Kolata, "A New Treatment's Tantalizing Promise Brings Heartbreaking Ups and Downs," *New York Times*, July 8, 2012, http://www.nytimes.com/2012/07 /09/health/new-frontiers-of-cancer-treatment-bring-breathtaking-swings.html.

27   due to change as the burgeoning field of pharmacogenetics: "One Size Does Not Fit All: The Promise of Pharmacogenomics," National Center for Biotechnology Information, Science Primer, revised March 31, 2004, http://www.ncbi.nlm.nih .gov/About/primer/pharm.html.

27   the "mobile health" revolution: "mHealth in the Developing World," m+Health, accessed October 23, 2012, http://mplushealth.com/en/SiteRoot/MHme/Overview /mHealth-in-the-Developing-World/.

28   Mobile phones are now used: Lakshminarayanan Subramanian et al., "SmartTrack," CATER (Cost-effective Appropriate Technologies for Emerging Region), New York University, accessed October 11, 2012, http://cater.cs.nyu.edu/smarttrack#ref3.

30   tiny microchip that uses low-radiation: Kevin Spak, "Coming Soon: X-Ray Phones," *Newser*, April 20, 2012, http://www.newser.com/story/144464/coming-soon-x-ray -phones.html.

30   how could a dog eat his cloud storage drive?: A *New Yorker* cartoon by Tom Cheney in 2012 expressed a similar idea. Its caption read "The Cloud Ate My Homework." See "Cartoons from the Issue," *New Yorker*, October 8, 2012, http://www.newyorker .com/humor/issuecartoons/2012/10/08/cartoons_20121001#slide=5.

CHAPTER 2   The Future of Identity, Citizenship and Reporting

35   While many worry about the phenomenon of confirmation bias: Eli Pariser describes this as a "filter bubble" in his book *The Filter Bubble: What the Internet Is Hiding from You* (New York: Penguin Press, 2011).

35   a recent Ohio State University study: R. Kelly Garrett and Paul Resnick, "Resisting Political Fragmentation on the Internet," *Daedalus* 140, no. 4 (Fall 2011): 108–120, doi:10.1162/DAED_a_00118.

37   famously dissected how ethnically popular names: Steven D. Levitt and Stephen J. Dubner, *Freakonomics: A Rogue Economist Explores the Hidden Side of Everything* (New York: William Morrow, 2005); their study showed that the names were not the cause of a child's success or failure, but a symptom of other indicators (particularly socioeconomic ones) that do influence a child's chances. See Steven D. Levitt and Stephen J. Dubner, "A Roshanda by Any Other Name," *Slate,* April 11, 2005, http://www.slate.com/articles/business/the_dismal_science/2005/04/a_roshanda _by_any_other_name.single.html.

38   Wall Street bankers hired: Nick Bilton, "Erasing the Digital Past," *New York Times,* April 1, 2011, http://www.nytimes.com/2011/04/03/fashion/03reputation .html?pagewanted=all.

40   Assange shared his two basic arguments on this subject: Julian Assange in discussion with the authors, June 2011.

43   lightning rod, as Assange called himself: Atika Shubert, "WikiLeaks Editor Julian Assange Dismisses Reports of Internal Strife," CNN, October 22, 2010, http:// articles.cnn.com/2010-10-22/us/wikileaks.interview_1_julian-assange-wikileaks -afghan-war-diary?_s=PM:US.

43   "Sources speak with their feet": Julian Assange in discussion with the authors, June 2011.

44   WikiLeaks lost its principal website URL: James Cowie, "WikiLeaks: Moving Target," *Renesys* (blog), December 7, 2010, http://www.renesys.com/blog/2010/12 /wikileaks-moving-target.shtml.

44   "mirror" sites: Ravi Somaiya, "Pro-Wikileaks Activists Abandon Amazon Cyber Attack," BBC, December 9, 2010, http://www.bbc.com/news/technology-11957367.

45   Alexei Navalny, a Russian blogger: Matthew Kaminski, "The Man Vladimir Putin Fears Most," *Wall Street Journal,* March 3, 2012, http://online.wsj.com/article /SB10001424052970203986604577257321601811092.html; "Russia Faces to Watch: Alexei Navalny," BBC, June 12, 2012, http://www.bbc.co.uk/news/world-europe -18408297.

45   donate toward its operating costs via PayPal: Tom Parfitt, "Alexei Navalny: Russia's New Rebel Who Has Vladimir Putin in His Sights," *Guardian* (Manchester), January 15, 2012, http://www.guardian.co.uk/theguardian/2012/jan/15/alexei-navalny -profile-vladimir-putin.

45   set of leaked documents: "Russia Checks Claims of $4bn Oil Pipeline Scam," BBC, November 17, 2010, http://www.bbc.co.uk/news/world-europe-11779154.

45   the Party of Crooks and Thieves: Tom Parfitt, "Russian Opposition Activist Alexei Navalny Fined for Suggesting United Russia Member Was Thief," *Telegraph* (London), June 5, 2012, http://www.telegraph.co.uk/news/worldnews/europe /russia/9312508/Russian-opposition-activist-Alexei-Navalny-fined-for-suggesting -United-Russia-member-was-thief.html; Stephen Ennis, "Profile: Russian Blogger

Alexei Navalny," BBC, August 7, 2012, http://www.bbc.co.uk/news/world-europe -16057045.

45 arrested, imprisoned, spied on and investigated for embezzlement: Ellen Barry, "Rousing Russia with a Phrase," *New York Times,* December 9, 2011, http://www.nytimes .com/2011/12/10/world/europe/the-saturday-profile-blogger-aleksei-navalny-rouses -russia.html. Robert Beckhusen, "Kremlin Wiretaps Dissident Blogger—Who Tweets the Bug," *Danger Room* (blog), *Wired,* August 8, 2012, http://www.wired .com/dangerroom/2012/08/navalny-wiretap/. "Navalny Charged with Embezzlement, Faces up to 10 Years," *RT* (Moscow), last updated August 1, 2012, http://rt .com/politics/navalny-charged-travel-ban-476/.

46 his name recognition: Parfitt, "Alexei Navalny: Russia's New Rebel Who Has Vladimir Putin in His Sights," http://www.guardian.co.uk/theguardian/2012/jan/15 /alexei-navalny-profile-vladimir-putin.

46 banned from appearing on state-run television: Kaminski, "The Man Vladimir Putin Fears Most," http://online.wsj.com/article/SB10001424052970203986604577257325732 1601811092.html.

46 Mikhail Khodorkovsky: "Mikhail Khodorkovsky," *New York Times,* last updated August 8, 2012, http://topics.nytimes.com/top/reference/timestopics/people/k /mikhail_b_khodorkovsky/index.html; Andrew E. Kramer, "Amid Political Prosecutions, Russian Court Issues Ruling Favorable to Oil Tycoon," *New York Times,* August 1, 2012, http://www.nytimes.com/2012/08/02/world/europe/russian-court -issues-favorable-ruling-to-oil-tycoon.html. At the time of the publication of this book, Khodorkovsky remained in prison. There was some speculation that President Vladimir Putin might commute the thirteen-year prison sentence.

46 Boris Berezovsky: Svetlana Kalmykova, "Oligarch Berezovsky Faces New Charges," *Voice of Russia* (Moscow), May 29, 2012, http://english.ruvr.ru/2012_05_29 /76399306/.

46 badly doctored photograph: "Russian Blogger Navalny Unmasks 'Kremlin' Photo Smear," BBC, January 10, 2012, http://www.bbc.co.uk/news/world-europe -16487469.

46 formally charging him with embezzlement: Ellen Barry, "Russia Charges Anticorruption Activist in Plan to Steal Timber," *New York Times,* July 31, 2012, http:// www.nytimes.com/2012/08/01/world/europe/aleksei-navalny-charged-with -embezzlement.html.

46 The charges, carrying a maximum sentence: Ibid.

46 150,000 Sony customer records released by the hacker group LulzSec in 2011: Mathew J. Schwartz, "Sony Hacked Again, 1 Million Passwords Exposed," *InformationWeek,* June 3, 2011, http://www.informationweek.com/security/attacks/sony -hacked-again-1-million-passwords-ex/229900111.

47 Assange told us he redacted only to reduce the international pressure: Julian Assange in discussion with the authors, June 2011.

47 "zero tolerance" approach: Charlie Savage, "Holder Directs U.S. Attorneys to Track Down Paths of Leaks," *New York Times,* June 8, 2012, http://www.nytimes .com/2012/06/09/us/politics/holder-directs-us-attorneys-to-investigate-leaks .html?pagewanted=all.

48 unknowingly live-tweeted the covert raid: Reed Stevenson, Reuters, "Sohaib Athar Captures Osama bin Laden Raid on Twitter," *Huffington Post,* first posted May 2, 2011, last updated July 2, 2011, http://www.huffingtonpost.com/2011/05/02/osama -bin-laden-raid-twitter-sohaib-athar_n_856187.html.

48   Among the tweets: Ibid.; Sohaib Athar, Twitter post, May 1, 2011, 12:58 a.m., https://twitter.com/ReallyVirtual/status/64780730286358528. (Five of the tweets Sohaib Athar sent the night of the bin Laden raid: 1) "Helicopter hovering above Abbottabad at 1AM (is a rare event)" (his first tweet on the matter). 2) "Go away helicopter—before I take out my giant swatter :-/." 3) "A huge window shaking bang here in Abbottabad Cantt. I hope its not the start of something nasty :-S." 4) "@mohcin the few people online at this time of the night are saying one of the copters was not Pakistani . . ." 5) "Since taliban (probably) don't have helicopters, and since they're saying it was not 'ours,' so must be a complicated situation #abbottabad." See Rik Myslewski, "Pakistani IT Admin Leaks bin Laden Raid on Twitter," *Register,* May 2, 2011, http://www.theregister.co.uk/2011/05/02/bin_laden_raid_tweeted/.

54   Connectivity is relatively low: See low mobile penetration of countries at the bottom of the Press Freedom Index such as Eritrea and North Korea in "Mobile-Cellular Telephone Subscriptions Per 100 Inhabitants," International Telecommunication Union (ITU), ICT Data and Statistics (IDS), accessed October 15, 2012, http://www.itu.int/ITUD/ict/statistics/, and "Press Freedom Index 2011/2012," Reporters Without Borders (RSF), accessed October 15, 2012, http://en.rsf.org/press-freedom-index-2011-2012,1043.html.

54   Warlords operating: "ICC/DRC: Second Trial of Congolese Warlords," Human Rights Watch, News, November 23, 2009, http://www.hrw.org/news/2009/11/23/iccdrc-second-trial-congolese-warlords; Marlise Simons, "International Criminal Court Issues First Sentence," *New York Times,* July 10, 2012, http://www.nytimes.com/2012/07/11/world/europe/international-criminal-court-issues-first-sentence.html.

55   Presidential Records Act: "Presidential Records Act (PRA) of 1978," National Archives, Presidential Libraries, Laws and Regulations, accessed October 12, 2012, http://www.archives.gov/presidential-libraries/laws/1978-act.html; "Presidential Records," National Archives, Basic Laws and Authorities, accessed October 12, 2012, http://www.archives.gov/about/laws/presidential-records.html.

56   Hamza Kashgari posted an imaginary conversation with the Prophet Muhammad: Mike Giglio, "Saudi Writer Hamza Kashgari Detained in Malaysia over Muhammad Tweets," *Daily Beast,* February 10, 2012, http://www.thedailybeast.com/articles/2012/02/08/twitter-aflame-with-fatwa-against-saudi-writer-hamza-kashgari.html.

56   deleted them within six hours of posting: Asma Alsharif and Amena Bakr, "Saudi Writer May Face Trial over Prophet Mohammad," Reuters, February 13, 2012, http://www.reuters.com/article/2012/02/13/us-saudi-blogger-idUSTRE81C13720120213.

56   creation of a Facebook group: Liz Gooch and J. David Goodman, "Malaysia Detains Saudi over Twitter Posts on Prophet," *New York Times,* February 10, 2012, http://www.nytimes.com/2012/02/11/world/asia/malaysia-detains-saudi-over-twitter-posts-on-prophet.html.

56   Kashgari fled to Malaysia but was deported: Ellen Knickmeyer, "Saudi Tweeter Is Arrested in Malaysia," *Wall Street Journal,* February 10, 2012, http://online.wsj.com/article/SB10001424052970204642604577213553613859184.html; Nadim Kawach, "Malaysia Deports Saudi over Twitter Posts," *Emirates 24/7,* February 11, 2012, http://www.emirates247.com/news/region/malaysia-deports-saudi-over-twitter-posts-2012-02-11-1.442363.

56   charges of blasphemy: "Saudi Writer Kashgari Deported," Freedom House, News and Updates, accessed October 12, 2012, http://www.freedomhouse.org/article/saudi-writer-kashgari-deported; "Saudi Arabia: Writer Faces Apostasy Trial,"

Human Rights Watch (HRW), News, February 13, 2012, http://www.hrw.org /news/2012/02/13/saudi-arabia-writer-faces-apostasy-trial.

56   a subsequent August 2012 apology: Laura Bashraheel, "Hamza Kashgari's Poem from Prison," *Saudi Gazette* (Jeddah), last updated Tuesday, August 21, 2012, http://www .saudigazette.com.sa/index.cfm?method=home.regcon&contentid=20120821133653.

59   murder of a prominent actress by a stalker: "The Drivers Privacy Protection Act (DPPA) and the Privacy of Your State Motor Vehicle Record," Electronic Privacy Information Center, accessed October 13, 2012, http://epic.org/privacy/drivers/.

59   leak of the late Judge Robert Bork's video-rental information: "Existing Federal Privacy Laws," Center for Democracy and Technology, accessed October 13, 2012, https://www.cdt.org/privacy/guide/protect/laws.php#vpp.

59   Texas lawsuit: "Harris v. Blockbuster," Electronic Privacy Information Center, accessed October 13, 2012, http://epic.org/amicus/blockbuster/default.html; Cathryn Elaine Harris, Mario Herrera, and Maryam Hosseiny v. Blockbuster, Inc., Settlement, District Court for the Northern District of Texas Dallas Division, Civil Action No. 3:09-cv-217-M, http://www.scribd.com/doc/28540910/Lane-v -Facebook-Blockbuster-Settlement.

61   Syrian opposition members and foreign aid workers: Ben Brumfield, "Computer Spyware Is Newest Weapon in Syrian Conflict," CNN, February 17, 2012, http:// articles.cnn.com/2012-02-17/tech/tech_web_computer-virus-syria_1_opposition -activists-computer-viruses-syrian-town?_s=PM:TECH.

61   Information technology (IT) specialists outside of Syria: Ibid.

61   One aid worker had downloaded a file: Ibid.

62   crash of a high-speed train in Wenzhou: "China Train Crash: Signal Design Flaw Blamed," BBC, July 28, 2011, http://www.bbc.co.uk/news/world-asia-pacific -14321060.

62   posts on *weibo*s: Michael Wines and Sharon LaFraniere, "In Baring Facts of Train Crash, Blogs Erode China Censorship," *New York Times,* July 28, 2011, http://www .nytimes.com/2011/07/29/world/asia/29china.html?pagewanted=all.

62   result of a design flaw: Sharon LaFraniere, "Design Flaws Cited in Deadly Train Crash in China," *New York Times,* December 28, 2011, http://www.nytimes.com /2011/12/29/world/asia/design-flaws-cited-in-china-train-crash.html; "China Bullet Train Crash 'Caused by Design Flaws,'" BBC, December 28, 2011, http://www.bbc .co.uk/news/world-asia-china-16345592.

62   government sent directives to the media shortly after the crash: David Bandurski, "History of High-Speed Propaganda Tells All," *China Media Project,* July 25, 2011, http://cmp.hku.hk/2011/07/25/14036/?utm_source=twitterfeed&utm_medium =twitter.

63   In Somalia, telecommunications companies: Abdinasir Mohamed and Sarah Childress, "Telecom Firms Thrive in Somalia Despite War, Shattered Economy," *Wall Street Journal,* May 11, 2010, http://online.wsj.com/article /SB10001424052748704608104575220570113266984.html.

67   "trespass to chattels" tort has in some cases already been applied to cyberspace: Eric J. Sinrod, "Perspective: A Cyberspace Update for Hoary Legal Doctrine," *CNET,* April 4, 2007, http://news.cnet.com/A-cyberspace-update-for-hoary-legal-doctrine/ 2010-1030_3-6172900.html.

69   using a mix of mobile money platforms and the traditional *"hawala"* money-transfer system: Andrew Quinn, "Cell Phones May Be New Tool vs. Somalia Famine,"

Reuters, September 21, 2011, Africa edition, http://af.reuters.com/article/topNews /idAFJOE78KooL20110921.

69   forged new opportunities: Sahra Abdi, "Mobile Transfers Save Money and Lives in Somalia," Reuters, March 3, 2010, http://www.reuters.com/article/2010/03/03/us -somalia-mobiles-idUSTRE6222BY20100303.

70   mobile adoption has vastly outpaced computer use: Compare mobile cellular subscriptions to Internet subscriptions in 2010 for countries such as Equatorial Guinea, Mali, Niger, etc., in "Mobile-Cellular Subscriptions" and "Fixed (Wired) Internet Subscriptions," International Telecommunication Union (ITU), ICT Data and Statistics (IDS), accessed October 13, 2012, http://www.itu.int/ITU-D/ict/statistics/.

70   many people treat their phones like stereo systems: Michael Byrne, "Inside the Cell Phone File Sharing Networks of Western Africa (Q+A)," *Motherboard,* January 3, 2012, http://motherboard.vice.com/2012/1/3/inside-the-cell-phone-file-sharing -networks-of-western-africa-q-a.

70   promise even richer wearable experiences: Dena Cassella, "What Is Augmented Reality (AR): Augmented Reality Defined, iPhone Augmented Reality Apps and Games and More," *Digital Trends,* November 3, 2009, http://www.digitaltrends .com/mobile/what-is-augmented-reality-iphone-apps-games-flash-yelp-android-ar -software-and-more/.

70   Project Glass: Babak Parviz, Steve Lee, Sebastian Thrun, "Project Glass," Google+, April 4, 2012, https://plus.google.com/+projectglass/posts; Nick Bilton, "Google Begins Testing Its Augmented-Reality Glasses," *Bits* (blog), *New York Times,* April 4, 2012, http://bits.blogs.nytimes.com/2012/04/04/google-begins-testing-its -augmented-reality-glasses/.

70   and similar devices from other companies are on the way: Todd Wasserman, "Apple Patent Hints at Google Glass Competitor," *Mashable,* July 5, 2012, http://mashable .com/2012/07/05/apple-patent-google-glass/; Molly McHugh, "Google Glasses Are Just the Beginning: Why Wearable Computing Is the Future," *Digital Trends,* July 6, 2012, http://www.digitaltrends.com/computing/google-glasses-are-just-the -beginning-why-wearable-computing-is-the-future/#ixzz29PI4PWK4.

72   introducing bills that would force communications services: Declan McCullagh, "FBI: We Need Wiretap-Ready Web Sites—Now," *CNET,* May 4, 2012, http://news .cnet.com/8301-1009_3-57428067-83/fbi-we-need-wiretap-ready-web-sites-now/; Charlie Savage, "As Online Communications Stymie Wiretaps, Lawmakers Debate Solutions," *New York Times,* February 17, 2011, http://www.nytimes.com/2011/02 /18/us/18wiretap.html.

72   Napster, was shut down: Matt Richtel, "Technology; Judge Orders Napster to Police Trading," *New York Times,* March 7, 2001, http://www.nytimes.com /2001/03/07/business/technology-judge-orders-napster-to-police-trading.html?ref =marilynhallpatel; Matt Richtel, "With Napster Down, Its Audience Fans Out," *New York Times,* July 20, 2001, http://www.nytimes.com/2001/07/20/business /technology-with-napster-down-its-audience-fans-out.html?pagewanted=all&src =pm.

72   capable of blocking the transfer of 99.4 percent of copyrighted material: Matt Richtel, "Napster Appeals an Order to Remain Closed Down," *New York Times,* July 13, 2001, http://www.nytimes.com/2001/07/13/business/technology-napster -appeals-an-order-to-remain-closed-down.html; Lawrence Lessig, *Free Culture: How Big Media Uses Technology and the Law to Lock Down Culture and Control Creativity*

(New York: Penguin Press, 2004), 73–74, http://www.free-culture.cc/freeculture .pdf.

72   Bluetooth-enabled phones to call and text complete strangers within range: "Beware: Dangers of Bluetooth in Saudi . . . ," *Emirates 24/7,* December 1, 2010, http://www .emirates247.com/news/region/beware-dangers-of-bluetooth-in-saudi-2010-12-01 -1.323699; Associated Press (AP), "In Saudi Arabia, a High-Tech Way to Flirt," MSNBC, August 11, 2005, http://www.msnbc.msn.com/id/8916890/ns/world _news-mideast_n_africa/t/saudi-arabia-high-tech-way-flirt/#.UJBUosVG-8A.

72   Etisalat sent nearly 150,000 of its BlackBerry users: Margaret Coker and Stuart Weinberg, "RIM Warns Update Has Spyware," *Wall Street Journal,* July 23, 2009, http://online.wsj.com/article/SB124827172417172239.html; John Timmer, "UAE Cellular Carrier Rolls Out Spyware as a 3G 'Update,'" *Ars Technica,* July 23, 2009, http://arstechnica.com/business/2009/07/mobile-carrier-rolls-out-spyware-as-a-3g -update/.

72   required update for "service enhancements": "UAE Spyware Blackberry Update," *Digital Trends,* July 22, 2009, http://www.digitaltrends.com/mobile/uae-spyware -blackberry-update/.

73   RIM, distanced itself: George Bevir, "Etisalat Accused in Surveillance Patch Fiasco," *Arabian Business,* July 21, 2009, http://www.arabianbusiness.com/etisalat-accused -in-surveillance-patch-fiasco-15698.html; see also, Adam Schreck, Associated Press (AP), "United Arab Emirates, Saudi Arabia to Block BlackBerry over Security Fears," *Huffington Post,* August 1, 2010, http://www.huffingtonpost.com/2010/08 /01/uae-saudi-arabia-blackberry-ban_n_666581.html.

73   the U.A.E. and its neighbor Saudi Arabia both called for bans: Margaret Coker, Tim Falconer, Phred Dvorak, "U.A.E. Puts the Squeeze on BlackBerry," *Wall Street Journal,* August 2, 2010, http://online.wsj.com/article/SB100014240527 4870470230457540249330069891 2.html; Kayla Webley, "UAE, Saudi Arabia Ban the Blackberry," *Time,* August 5, 2010, http://www.time.com/time/specials/packages /article/0,28804,2008434_2008436_2008440,00.html; "Saudi Arabia Begins Black-berry Ban, Users Say," BBC, August 6, 2010, http://www.bbc.co.uk/news/world -middle-east-10888954.

73   India chimed in: Bappa Majumdar and Devidutta Tripathy, "Setback for BlackBerry in India; Saudi Deal Seen," Reuters, August 11, 2010, India edition, http://in.reuters .com/article/2010/08/11/idINIndia-50769520100811.

73   resulted in five deaths: Laura Davis, "The Debate: Could the Behaviour Seen at the Riots Ever Be Justified?," *Notebook* (blog), *Independent* (London), August 8, 2012, http://blogs.independent.co.uk/2012/08/08/the-debate-could-the-behaviour-seen -at-the-riots-ever-be-justified/.

73   estimated £300 million ($475 million) in property damage: John Benyon, "Eng-land's Urban Disorder: The 2011 Riots," *Political Insight,* March 28, 2012, http:// www.politicalinsightmagazine.com/?p=911; "A Little Bit of History Repeating," *Inside Housing,* July 27, 2012, http://www.insidehousing.co.uk/tenancies/a-little-bit -of-history-repeating/6522947.article.

73   called on BlackBerry to suspend its messaging service: Sky News Newsdesk, Twitter post, August 9, 2011, 5:32 a.m., https://twitter.com/SkyNewsBreak/sta-tus/100907315603054592; Bill Ray, "Tottenham MP Calls for BlackBerry Messeng-ing Suspension," *Register,* August 9, 2011, http://www.theregister.co.uk/2011/08/09 /bbm_suspension/.

73   "when we know [people] are plotting violence": "PM Statement on Disorder in En-

gland," Number 10 (official website of the British Prime Minister's Office), August 11, 2011, http://www.number10.gov.uk/news/pm-statement-on-disorder-in-england/.

73 "give the police the technology": Rich Trenholm, "Cameron Considers Blocking Twitter, Facebook, BBM after Riots," *CNET,* August 11, 2011, http://crave.cnet .co.uk/software/cameron-considers-blocking-twitter-facebook-bbm-after-riots -50004693/; Olivia Solon, "Cameron Suggests Blocking Potential Criminals from Social Media," *Wired UK,* August 11, 2011, http://www.wired.co.uk/news/archive /2011-08/11/david-cameron-social-media.

73 industry cooperation with law enforcement was sufficient: "Social Media Talks About Rioting 'Constructive,'" BBC, August 25, 2011, http://www.bbc.co.uk/news /uk-14657456.

74 Bitcoins: Bitcoin is the most successful experiment in digital currency today; it uses a mix of peer-to-peer networking and cryptographic signatures to process online payments. The value of the currency has fluctuated wildly since its inception; the first publicly traded Bitcoins went for 3 cents, and a little more than a year later they were valued at $29.57 apiece. Bitcoins are held in digital "wallets," and are used to pay for a wide range of virtual and physical goods. At the illicit online market called the Silk Road, where people can use encrypted channels to buy illegal drugs, Bitcoins are the sole currency and generate approximately $22 million in annual sales, according to a recent study. See Andy Greenberg, "Black Market Drug Site 'Silk Road' Booming: $22 Million in Annual Sales," *Forbes,* August 6, 2012, http://www.forbes.com/sites /andygreenberg/2012/08/06/black-market-drug-site-silk-road-booming-22-million -in-annual-mostly-illegal-sales/; Nicolas Christin, "Traveling the Silk Road: A Measurement Analysis of a Large Anonymous Online Marketplace" (working paper, INI/CyLab, Carnegie Mellon, Pittsburgh, PA, August 1, 2012), http://arxiv.org/pdf /1207.7139v1.pdf.

74 "there is no clear mechanism": Bruno Ferrari in discussion with the authors, November 2011.

75 not democratic or democratic in name only: Arch Puddington, *Freedom in the World 2012: The Arab Uprisings and Their Global Repercussions,* Freedom House, accessed October 15, 2012, http://www.freedomhouse.org/sites/default/files/FIW%202012% 20Booklet_0.pdf.

75 among the least connected societies in the world: See low percentages of mobile phone and/or Internet users of countries considered to be among the world's most repressive societies, such as Equatorial Guinea, Eritrea and North Korea, in *Worst of the Worst 2012: The World's Most Repressive Societies,* Freedom House, accessed October 15 2012, http://www.freedomhouse.org/sites/default/files/Worst%20of%20 the%20Worst%202012%20final%20report.pdf, "Mobile-Cellular Telephone Subscriptions Per 100 Inhabitants" and "Percentage of Individuals Using the Internet," International Telecommunication Union (ITU), ICT Data and Statistics (IDS), accessed October 15, 2012, http://www.itu.int/ITU-D/ict/statistics/.

75 "Today's dictators and authoritarians are far more sophisticated": William J. Dobson, *The Dictator's Learning Curve: Inside the Global Battle for Democracy* (New York: Doubleday, 2012), 4.

75 Dobson identifies numerous avenues: Ibid.

76 "conscious, man-made projects": Ibid., 8.

76 the world's autocracies will go: See low Internet penetration rates of countries considered to be among the world's most repressive societies, such as Equatorial Guinea, Eritrea and North Korea, in *Worst of the Worst 2012: The World's Most Repressive*

Societies, Freedom House, accessed October 15, 2012, http://www.freedomhouse.org /sites/default/files/Worst%20of%20the%20Worst%202012%20final%20report .pdf, and "Percentage of Individuals Using the Internet," International Telecom- munication Union (ITU), ICT Data and Statistics (IDS), accessed October 15, 2012, http://www.itu.int/ITU-D/ict/statistics/.

77   A team at Carnegie Mellon demonstrated in a 2011 study: Alessandro Acquisti, Ralph Gross, Fred Stutzman, "Faces of Facebook: Privacy in the Age of Augmented Reality," Heinz College and CyLab, Carnegie Mellon University (presented at the 2011 Black Hat security conference, Las Vegas, NV, August 3–4, 2011), http://media .blackhat.com/bh-us-11/Acquisti/BH_US_11_Acquisti_Faces_of_Facebook_Slides .pdf.; Declan McCullagh, "Face-Matching with Facebook Profiles: How It Was Done," *CNET,* August 4, 2011, http://news.cnet.com/8301-31921_3-20088456-281 /face-matching-with-facebook-profiles-how-it-was-done/.

78   Constituted in 2009: "UIDAI Background," Unique Identification Authority of India, accessed October 13, 2012, http://uidai.gov.in/about-uidai.html.

78   collectively called Aadhaar (meaning "foundation" or "support"): "Aadhaar Con- cept," Unique Identification Authority of India, accessed October 13, 2012, http:// uidai.gov.in/aadhaar.html.

78   unique twelve-digit identity: "What Is Aadhaar?," Unique Identification Author- ity of India, accessed October 13, 2012, http://uidai.gov.in/what-is-aadhaar-number .html.

78   a person's biometric data, including fingerprints and iris scans: Sunil Dabir and Umesh Ujgare, "Aadhaar: The Numbers for Life," *News on Air* (New Delhi), accessed October 13, 2012, http://www.newsonair.nic.in/AADHAAR-UID-Card -THE-NUMBERS-FOR-LIFE.asp.

79   bank account that is tied to his or her UID number: Surabhi Agarwal and Remya Nair, "UID-Enabled Bank Accounts in 2–3 Months," *Mint with the Wall Street Journal* (New Delhi), May 17, 2011, http://www.livemint.com/Politics /Go6diBWitIaus61Xud70EK/UIDenabled-bank-accounts-in-23-months.html; "Reform by Numbers," *Economist,* January 14, 2012, http://www.economist.com /node/21542814.

79   less than 3 percent of the Indian population is registered to pay income tax: "Salaried Taxpayers May Be Spared Filing Returns," *Business Standard* (New Delhi), Janu- ary 19, 2011, http://business-standard.com/india/news/salaried-taxpayers-may-be -spared-filing-returns/422225/.

79   Identity Cards Act of 2006: "Identity Cards Act 2006," The National Archives (United Kingdom), Browse Legislation, accessed October 15, 2012, http://www .legislation.gov.uk/ukpga/2006/15/introduction.

79   Britain's newly elected coalition government scrapped the plan in 2010: Alan Tra- vis, "ID Cards Scheme to Be Scrapped Within 100 Days," *Guardian* (Manchester), May 27, 2010, http://www.guardian.co.uk/politics/2010/may/27/theresa-may -scrapping-id-cards; "Identity Cards Scheme Will Be Axed 'Within 100 Days,'" BBC, May 27, 2010, http://news.bbc.co.uk/2/hi/8707355.stm.

80   States must get the full and informed consent: "Opinion 15/2011 on the Definition of Consent," Article 29 Data Protection Working Party, European Commission, adopted July 13, 2011, http://ec.europa.eu/justice/policies/privacy/docs/wpdocs /2011/wp187_en.pdf.

80   Member states are further required: "EU Directive 95/46/EC—The Data Protection

Directive: Chapter III Judicial Remedies, Liability and Sanctions," Data Protection Commissioner, http://www.dataprotection.ie/viewdoc.asp?DocID=94.

CHAPTER 3 The Future of States

84   YouTube in Iran: Gwen Ackerman and Ladane Nasseri, "Google Confirms Gmail and YouTube Blocked in Iran Since Feb. 10," *Bloomberg,* February 13, 2012, http://www.bloomberg.com/news/2012-02-13/google-confirms-gmail-and-youtube-blocked-in-iran-since-feb-10.html.

85   We recommend the 2006 book *Who Controls the Internet?:* Jack Goldsmith and Tim Wu, *Who Controls the Internet?: Illusions of a Borderless World* (New York: Oxford University Press, 2006).

85   most users tend to stay within their own cultural spheres: Author's determination based on ten years as CEO of Google and two as executive chairman.

86   Particular terms like "Falun Gong": Mark McDonald, "Watch Your Language! (In China, They Really Do)," *Rendezvous* (blog), *International Herald Tribune,* the global edition of the *New York Times,* March 13, 2012, http://rendezvous.blogs.nytimes.com/2012/03/13/watch-your-language-and-in-china-they-do/.

86   following a contentious trip: Observations from Google's executive chairman, Eric Schmidt.

86   Chinese officials had hired nearly three hundred thousand: Nate Anderson, "280,000 Pro-China Astroturfers Are Running Amok Online," *Ars Technica,* March 26, 2010, http://arstechnica.com/tech-policy/news/2010/03/280000-pro-china-astroturfers-are-running-amok-online.ars; Rebecca MacKinnon, "China, the Internet, and Google," prepared remarks (not delivered) for Congressional-Executive Commission on China, March 1, 2010, http://rconversation.blogs.com/MacKinnonCECC_Mar1.pdf; David Bandurski, "China's Guerrilla War for the Web," *Far Eastern Economic Review,* July 2008, http://www.feer.com/essays/2008/august/chinas-guerrilla-war-for-the-web. Note: the 280,000 figure was originally published in 2008, but restated in 2010.

86   In a white paper released in 2010: Full Text: *The Internet in China, IV. Basic Principles and Practices of Internet Administration* (June 8, 2010), Chinese Government's Official Web Portal, http://english.gov.cn/2010-06/08/content_1622956_6.htm.

87   YouTube was blocked: Tom Zeller, Jr., "YouTube Banned in Turkey after Insults to Ataturk," *The Lede* (blog), *New York Times,* March 7, 2007, http://thelede.blogs.nytimes.com/2007/03/07/youtube-banned-in-turkey-after-insults-to-ataturk/.

87   YouTube agreed to block the videos: Jeffrey Rosen, "Google's Gatekeepers," *New York Times Magazine,* November 28, 2008, http://www.nytimes.com/2008/11/30/magazine/30google-t.html?partner=permalink&exprod=permalink.

87   some eight thousand websites: Ayla Albayrak, "Turkey Dials Back Plan to Expand Censorship," *Wall Street Journal,* August 6, 2011, http://online.wsj.com/article/SB10001424053111903885604576490253692671470.html.

87   four-tier system of censorship: Sebnem Arsu, "Internet Filters Set Off Protests Around Turkey," *New York Times,* May 15, 2011, http://www.nytimes.com/2011/05/16/world/europe/16turkey.html?_r=3&.

88   thousands of people in more than thirty cities: Ibid.

88   Under pressure, the government dialed back its plan: Ayla Albayrak, "Turkey Dials Back Plan to Expand Censorship," *Wall Street Journal,* August 6, 2011.

88   more aggressive filtering framework: "New Internet Filtering System Condemned as Backdoor Censorship," Reporters Without Borders, December 2, 2011, http://en.rsf .org/turquie-new-internet-filtering-system-02-12-2011,41498.html.

88   Reporters Without Borders: Ibid.

88   When a Turkish newspaper reported: "Internet Filters Block Evolution Website for Children in Turkey," *Hurriyet* (Istanbul), December 8, 2011, http://www .hurriyetdailynews.com/internet-filters-block-evolution-website-for-children-in -turkey.aspx?pageID=238&nID=8709&NewsCatID=374; Sara Reardon, "Controversial Turkish Internet Censorship Program Targets Evolution Sites," *Science,* December 9, 2011, http://news.sciencemag.org/scienceinsider/2011/12/controversial -turkish-internet-c.html.

89   In South Korea, for example, the National Security Law: "Countries Under Surveillance: South Korea," Reporters Without Borders, accessed October 21, 2012, http:// en.rsf.org/surveillance-south-korea,39757.html.

89   government blocked some forty websites: Ibid.

89   took down a dozen accounts: Lee Tae-hoon, "Censorship on Pro-NK Websites Tight," *Korea Times,* September 9, 2010, http://www.koreatimes.co.kr/www/news /nation/2010/12/113_72788.html.

89   government blocks websites within Germany: "Europe," OpenNet Initiative, accessed October 21, 2012, http://opennet.net/research/regions/europe; "Germany," OpenNet Initiative, accessed October 21, 2012, http://opennet.net/research/profiles /germany.

89   despite promising its citizens: Clara Chooi, "Najib Repeats Promise of No Internet Censorship," *Malaysian Insider* (Kuala Lumpur), April 24, 2011, http://www .themalaysianinsider.com/malaysia/article/najib-repeats-promise-of-no-internet -censorship.

89   codify it in its Bill of Guarantees: "Benefits," MSC Malaysia, accessed October 21, 2012, http://www.mscmalaysia.my/why_msc_malaysia.

89   blocked access to file-sharing sites: Ricky Laishram, "Malayasian Government Blocks the Pirate Bay, MegaUpload and Other File Sharing Websites," *Techie Buzz,* June 9, 2011, http://techie-buzz.com/tech-news/malayasian-government-blocks -websites.html.

89   Malaysian Communications and Multimedia Commission: Wong Pek Mei, "MCMC Wants Block of 10 Websites That Allow Illegal Movie Downloads," *The Star* (Petaling Jaya), June 10, 2011, http://thestar.com.my/news/story.asp?file=/2011/6/10 /nation/20110610161330&sec=nation.

92   "We respect that each country has chosen for itself": Sukhbaatar Batbold (former prime minister of Mongolia) in discussion with the authors, November 2011.

93   Chile became the first country in the world: Tim Stevens, "Chile Becomes First Country to Guarantee Net Neutrality, We Start Thinking About Moving," *Engadget,* July 15, 2010, http://www.engadget.com/2010/07/15/chile-becomes-first-country-to -guarantee-net-neutrality-we-star/.

93   About half of Chile's 17 million people: See population in 2011 and percentage of Internet users in 2011 in "Midyear Population and Density—Custom Region— Chile, 2011," U.S. Census Bureau, International Data Base, accessed October 21, 2012, http://www.census.gov/population/international/data/idb/informationGateway .php and "Percentage of Individuals Using the Internet," International Telecommu-

nication Union (ITU), ICT Data and Statistics (IDS), accessed October 21, 2012, http://www.itu.int/ITU-D/ict/statistics/.

95   "halal Internet": Neal Ungerleider, "Iran Cracking Down Online with 'Halal Internet,'" *Fast Company,* April 18, 2011, http://www.fastcompany.com/1748123/iran-cracking-down-online-halal-internet.

95   official launch was imminent: Neal Ungerleider, "Iran's 'Second Internet' Rivals Censorship of China's 'Great Firewall,'" *Fast Company,* February 23, 2012, http://www.fastcompany.com/1819375/irans-second-internet-rivals-censorship-chinas-great-firewall.

95   "government-approved videos": David Murphy, "Iran Launches 'Mehr,' Its Own YouTube-like Video Hub," http://www.pcmag.com/article2/0,2817,2413014,00.asp.

95   the first phase the national "clean" Internet: Christopher Rhoads and Farnaz Fassihi, "Iran Vows to Unplug Internet," *Wall Street Journal,* updated December 19, 2011, http://online.wsj.com/article/SB10001424052748704889404576277391449900 2016.html; Nick Meo, "Iran Planning to Cut Internet Access to Rest of World," *Telegraph* (London), April 28, 2012, http://www.telegraph.co.uk/news/worldnews/middleeast/iran/9233390/Iran-planning-to-cut-internet-access-to-rest-of-world.html.

95   2012 ban on the import of foreign computer security software: S. Isayev and T. Jafarov, "Iran Bans Import of Foreign Computer Security Software," *Trend,* February 20, 2012, http://en.trend.az/regions/iran/1994160.html.

95   Iran's head of economic affairs told the country's state-run news agency: Rhoads and Fassihi, "Iran Vows to Unplug Internet," http://online.wsj.com/article/SB1000 1424052748704889404576277391449002016.html.

95   Pakistan has pledged to build something similar: "Request for Proposal: National URL Filtering and Blocking System," National ICT R&D Fund, accessed October 21, 2012, http://ictrdf.org.pk/RFP-%20URL%20Filtering%20%26%20 Blocking.pdf; Ungerleider, "Iran's 'Second Internet' Rivals Censorship of China's 'Great Firewall,'" http://www.fastcompany.com/1819375/irans-second-internet-rivals-censorship-chinas-great-firewall; Danny O'Brien, "Pakistan's Excessive Internet Censorship Plans," Committee to Protect Journalists (CPJ), March 1, 2012, http://www.cpj.org/internet/2012/03/pakistans-excessive-net-censorship-plans.php. It is worth noting that at the time of writing, the Pakistani program had been "shelved." See Shahbaz Rana, "IT Ministry Shelves Plan to Install Massive URL Blocking System," *The Express Tribune* (Karachi) (blog) with the *International Herald Tribune,* March 19, 2012, http://tribune.com.pk/story/352172/it-ministry-shelves-plan-to-install-massive-url-blocking-system/.

96   company that owns 75 percent of North Korea's only official mobile network, Koryolink: "Mobile Phones in North Korea: Also Available to Earthlings," *Economist,* February 11, 2012, http://www.economist.com/node/21547295.

96   For North Korean subscribers, Koryolink service is a walled garden: Ibid.

97   North Korean daily *Rodong Sinmun* sending users the latest news by text message: Ibid.

97   pay their phone bills in euros: Ibid; David Matthew, "Understanding the Growth of KoryoLink," *NK News,* December 15, 2011, http://www.nknews.org/2011/12/understanding-koryo-link/.

97   leaping from three hundred thousand subscribers: "Mobile Phones in North Korea: Also Available to Earthlings," *Economist,* February 11, 2012.

97   Koryolink's gross operating margin: Ibid.

97   Ericsson and Nokia Siemens Networks: Steve Stecklow, Farnaz Fassihi and Loretta

Chao, "Chinese Tech Giant Aids Iran," *Wall Street Journal,* October 27, 2011, http:// online.wsj.com/article/SB10001424052970204644504576651503577823210.html? _nocache=1346874829284&user=welcome&mg=id-wsj.

97   Huawei actively promoted its products: Ibid.

97   Zaeim Electronic Industries Co., is also the favorite: Ibid.

97   Huawei claims to offer Zaeim only "commercial public-use products and services": Ibid.

97   Huawei published a press release: Huawei, "Statement Regarding Inaccurate and Misleading Claims About Huawei's Commercial Operations in Iran," press release, November 4, 2011, http://www.huawei.com/en/about-huawei/newsroom/press -release/hw-104191.htm.

97   "voluntarily restrict" its business operations: Huawei, "Statement Regarding Huawei's Commercial Operations in Iran," press release, December 9, 2011, http:// www.huawei.com/en/about-huawei/newsroom/press-release/hw-104866-statement -commercialoperations.htm.

100   "fully implement all of the intellectual property laws": Hu Jintao (former president of China) in discussion with a small group of business leaders at the Asia-Pacific Economic Cooperation (APEC) CEO Summit in 2011.

100   It's estimated that U.S. companies lost approximately $3.5 billion in 2009: *2010 Report to Congress on China's WTO Compliance,* United States Trade Representative (December 2010), 5, http://www.ustr.gov/webfm_send/2460.

100   79 percent of all copyright-infringing goods: Ibid., 92.

100   Russia, India and Pakistan have all been singled out: *2011 Special 301 Report,* United States Trade Representative, see "Section II: Country Reports Priority Watch List," 25, 28, 30, http://www.ustr.gov/webfm_send/2841.

100   Israel and Canada: Ibid., 27, 29.

103   definition of cyber warfare offered by the former U.S. counterterrorism chief Richard Clarke: Richard A. Clarke and Robert K. Knake, *Cyber War: The Next Threat to National Security and What to Do About It* (New York: Ecco, 2010), 6.

104   In October 2012, the U.S. secretary of defense, Leon Panetta, warned: Elisabeth Bumiller and Thom Shanker, "Panetta Warns of Dire Threat of Cyberattack on U.S.," *New York Times,* October 11, 2012, http://www.nytimes.com/2012/10/12 /world/panetta-warns-of-dire-threat-of-cyberattack.html?hp&_r=1&.

104   "war as a continuation of policy by other means": Carl von Clausewitz, *On War* (Baltimore: Penguin Books, 1968). The original quote is "war as a continuation of politik by other means."

105   "it's just much harder to know who took the shot at you": Craig Mundie in discussion with the authors, November 2011.

105   Mundie calls cyber-espionage tactics "weapons of mass disruption": Craig Mundie, "Information Security in the Digital Decade." Remarks at the American Chamber of Commerce in Bangkok, Thailand, October 20, 2003, http://www.microsoft.com /en-us/news/exec/craig/10-20security.aspx.

105   until a virus known as Flame, discovered in 2012, claimed that title: "Resource 207: Kaspersky Lab Research Proves That Stuxnet and Flame Developers Are Connected," Kaspersky Lab, June 11, 2012, http://www.kaspersky.com/about/news /virus/2012/Resource_207_Kaspersky_Lab_Research_Proves_that_Stuxnet_and _Flame_Developers_are_Connected.

105   causing the centrifuges to abruptly speed up or slow down: David E. Sanger, "Obama Order Sped Up Wave of Cyberattacks Against Iran," *New York Times,* June 1, 2012,

http://www.nytimes.com/2012/06/01/world/middleeast/obama-ordered-wave-of
-cyberattacks-against-iran.html?_r=1&ref=davidesanger&pagewanted=all.

106   perhaps unwittingly introduced by a Natanz employee on a USB flash drive: Ibid.

106   as the Iranian president, Mahmoud Ahmadinejad, admitted: Julian Borger and Saeed Kamali Dehghan, "Attack on Iranian Nuclear Scientists Prompts Hit Squad Claims," *Guardian* (Manchester), November 29, 2010, http://www.guardian.co.uk /world/2010/nov/29/iranian-nuclear-scientists-attack-claims.

106   had escaped "into the wild": Sanger, "Obama Order Sped Up Wave of Cyber-attacks Against Iran," http://www.nytimes.com/2012/06/01/world/middleeast /obama-ordered-wave-of-cyberattacks-against-iran.html?_r=1&ref=davides anger&pagewanted=all.

106   references to dates and biblical stories: Elinor Mills, "Stuxnet: Fact vs. Theory," *CNET,* October 5, 2010, http://news.cnet.com/8301-27080_3-20018530-245.html.

106   written by as many as thirty people: Michael Joseph Gross, "A Declaration of Cyber-War," *Vanity Fair,* April 2011, http://www.vanityfair.com/culture/features/2011/04 /stuxnet-201104.

106   an early variant of Stuxnet: Elinor Mills, "Shared Code Indicates Flame, Stuxnet Creators Worked Together," *CNET,* June 11, 2012, http://news.cnet.com/8301-1009 _3-57450292-83/shared-code-indicates-flame-stuxnet-creators-worked-together/.

106   Unnamed Obama administration officials confirmed: Sanger, "Obama Order Sped Up Wave of Cyberattacks Against Iran," http://www.nytimes.com/2012/06 /01/world/middleeast/obama-ordered-wave-of-cyberattacks-against-iran.html?_r =1&ref=davidesanger&pagewanted=all.

106   "Do you really expect me to *tell* you?": Meir Dagan in discussion with the authors, June 2012.

106   Olympic Games, was carried into the next administration: Sanger, "Obama Order Sped Up Wave of Cyberattacks Against Iran," http://www.nytimes.com/2012/06 /01/world/middleeast/obama-ordered-wave-of-cyberattacks-against-iran.html?_r =1&ref=davidesanger&pagewanted=all.

106   After building the malware and testing it: Ibid.

107   Larry Constantine . . . challenges Sanger's analysis: Larry Constantine, interview by Steven Cherry, "Stuxnet: Leaks or Lies?," *Techwise Conversations* (podcast), *IEEE Spectrum,* September 4, 2012, http://spectrum.ieee.org/podcast/computing /embedded-systems/stuxnet-leaks-or-lies.

107   Michael V. Hayden, the former CIA director, told Sanger: Sanger, "Obama Order Sped Up Wave of Cyberattacks Against Iran," http://www.nytimes.com/2012/06 /01/world/middleeast/obama-ordered-wave-of-cyberattacks-against-iran.html?_r =1&ref=davidesanger&pagewanted=all.

107   Sanger reported that American officials denied that Flame was part of the Olympic Games: Ibid.

107   security experts at Kaspersky Lab: "Resource 207: Kaspersky Lab Research Proves That Stuxnet and Flame Developers Are Connected," http://www.kaspersky.com /about/news/virus/2012/Resource_207_Kaspersky_Lab_Research_Proves_that _Stuxnet_and_Flame_Developers_are_Connected; Mills, "Shared Code Indicates Flame, Stuxnet Creators Worked Together," http://news.cnet.com/8301-1009_3 -57450292-83/shared-code-indicates-flame-stuxnet-creators-worked-together/.

107   identified a particular module, known as Resource 207: "Resource 207: Kaspersky Lab Research Proves That Stuxnet and Flame Developers Are Connected,"

http://www.kaspersky.com/about/news/virus/2012/Resource_207_Kaspersky_Lab
_Research_Proves_that_Stuxnet_and_Flame_Developers_are_Connected.

107 a senior Kaspersky researcher explained: Mills, "Shared Code Indicates Flame, Stuxnet Creators Worked Together," http://news.cnet.com/8301-1009_3-57450292-83
/shared-code-indicates-flame-stuxnet-creators-worked-together/.

108 diplomatic fight in 2007 over the Estonian government's decision: "Bronze Soldier
Installed at Tallinn Military Cemetery," *RIA Novosti* (Moscow), April 30, 2007,
http://en.rian.ru/world/20070430/64692507.html.

108 mass of prominent Estonian websites: Ian Traynor, "Russia Accused of Unleashing
Cyberwar to Disable Estonia," *Guardian* (Manchester), May 16, 2007, http://www
.guardian.co.uk/world/2007/may/17/topstories3.russia.

108 Estonia is often called the most wired country on Earth: Joshua Davis, "Hackers
Take Down the Most Wired Country in Europe," *Wired,* August 21, 2007, http://
www.wired.com/politics/security/magazine/15-09/ff_estonia?currentPage=all.

108 Urmas Paet, accused the Kremlin directly: Doug Bernard, "New Alarm Bells, and
Old Questions, About the Flame Virus and Cyber-War," *VOA* (blog), May 30, 2012,
http://blogs.voanews.com/digital-frontiers/tag/cyber-war/.

108 NATO and European Commission experts were unable to find evidence: "Estonia
Has No Evidence of Kremlin Involvement in Cyber Attacks," *RIA Novosti* (Moscow), June 9, 2007, http://en.rian.ru/world/20070906/76959190.html.

108 websites for the Georgian military and government were brought down: John Markoff, "Georgia Takes a Beating in the Cyberwar with Russia," *Bits* (blog), *New York
Times,* August 11, 2008, http://bits.blogs.nytimes.com/2008/08/11/georgia-takes-a
-beating-in-the-cyberwar-with-russia/; John Markoff, "Before the Gunfire, Cyberattacks," *New York Times,* August 12, 2008, http://www.nytimes.com/2008/08/13
/technology/13cyber.html.

108 Russian hackers targeted the Internet providers in Kyrgyzstan: Gregg Keizer, "Russian 'Cybermilitia' Knocks Kyrgyzstan Offline," *Computerworld,* January 28, 2009,
http://www.computerworld.com/s/article/9126947/Russian_cybermilitia_knocks
_Kyrgyzstan_offline.

108 shutting down 80 percent of the country's bandwidth for days: Christopher Rhoads,
"Kyrgyzstan Knocked Offline," *Wall Street Journal,* January 28, 2009, http://online
.wsj.com/article/SB123310906904622741.html.

108 Some believe the attacks were intended: Ibid.; "Kyrgyzstan to Close US Airbase,
Washington Says No Plans Made," *Hurriyet* (Istanbul), January 17, 2009, http://
www.hurriyet.com.tr/english/world/10796846.asp?scr=1.

109 In late 2009, Google detected unusual traffic within its network: David Drummond,
"A New Approach to China," *Google Blog,* January 12, 2010, http://googleblog
.blogspot.com/2010/01/new-approach-to-china.html.

109 Google's decision to alter its business position in China: David Drummond, "A New
Approach to China, an Update," *Google Blog,* March 22, 2010, http://googleblog
.blogspot.com/2010/01/new-approach-to-china.html.

109 Pentagon gave the directive to establish United States Cyber Command (USCYBER-
COM): "U.S. Cyber Command," U.S. Strategic Command, updated December
2011, http://www.stratcom.mil/factsheets/cyber_command/.

109 Robert Gates declared cyberspace to be the "fifth domain" of military operations: Misha Glenny, "Who Controls the Internet?," *Financial Times Magazine*
(London), October 8, 2010, http://www.ft.com/cms/s/2/3e52897c-d0ee-11df-a426
-00144feabdc0.html#axzz1nYp7grM6; Susan P. Crawford, "When We Wage Cyber-

war, the Whole Web Suffers," *Bloomberg,* April 25, 2012, http://www.bloomberg
.com/news/2012-04-25/when-we-wage-cyberwar-the-whole-web-suffers.html.

110 new "cyber-industrial complex" somewhere between $80 billion and $150 billion
annually: Ron Deibert and Rafal Rohozinski, "The New Cyber Military-Industrial
Complex," *Globe and Mail* (Toronto), March 28, 2011, http://www.theglobeandmail
.com/commentary/the-new-cyber-military-industrial-complex/article573990.

110 A raid on the Egyptian state security building after the country's: Ibid.; Eli Lake,
"British Firm Offered Spy Software to Egypt," *Washington Times,* April 25, 2011,
http://www.washingtontimes.com/news/2011/apr/25/british-firm-offered-spy
-software-to-egypt/?page=all#pagebreak.

111 Chinese telecom was contacted: WikiLeaks cable, "Subject: STIFLED POTENTIAL:
FIBER-OPTIC CABLE LANDS IN TANZANIA, Origin: Embassy Dar Es Salaam
(Tanzania), Cable time: Fri. 4 Sep 2009 04:48 UTC," http://www.cablegatesearch
.net/cable.php?id=09DARESSALAAM585.

111 Sichuan Hongda announced: Fumbuka Ng'wanakilala, "China Co Signs $3 Bln
Tanzania Coal, Iron Deal," Reuters, September 22, 2011, http://www.reuters.com
/article/2011/09/22/tanzania-china-mining-idUSL5E7KM1HU20110922.

111 loan agreement with China: "China, Tanzania Sign $1 Bln Gas Pipeline Deal:
Report," Reuters, September 30, 2011, Africa edition, http://af.reuters.com/article
/investingNews/idAFJOE78T08T20110930?pageNumber=1&virtualBrand
Channel=0.

111 State-owned enterprises make up 80 percent: "Emerging-Market Multinationals:
The Rise of State Capitalism," *Economist,* January 21, 2012, http://www.economist
.com/node/21543160.

111 $150 million loan for Ghana's e-governance venture: Andrea Marshall, "China's
Mighty Telecom Footprint in Africa," *eLearning Africa News Portal,* February 21, 2011,
http://www.elearning-africa.com/eLA_Newsportal/china%E2%80%99s-mighty
-telecom-footprint-in-africa/.

111 research hospital in Kenya: "East Africa: Kenya, China in Sh8 Billion University
Hospital Deal," *AllAfrica,* April 22, 2011, http://allafrica.com/stories/201104250544
.html.

111 "African Technological City" in Khartoum: John G. Whitesides, "Better Diplomacy, Better Science," *China Economic Review,* January 1, 1970, http://www
.chinaeconomicreview.com/content/better-diplomacy-better-science.

111 There are currently four main manufacturers: Opinion of the authors.

112 Some refer to this as the upcoming Code War: Michael Riley and Ashlee Vance,
"Cyber Weapons: The New Arms Race," *Bloomberg BusinessWeek,* July 20, 2011,
http://www.businessweek.com/magazine/cyber-weapons-the-new-arms-race
-07212011.html. As you can see, we did not coin the term "code war."

114 DDoS attacks crippled major government websites: Kim Zetter, "Lawmaker Wants
'Show of Force' Against North Korea for Website Attacks," *Wired,* July 10, 2009,
http://www.wired.com/threatlevel/2009/07/show-of-force/.

114 suggested that the network of attacking computers, or botnet, began in North Korea:
Choe Sang-Hun and John Markoff, "Cyberattacks Jam Government and Commercial Web Sites in U.S. and South Korea," *New York Times,* July 9, 2009, http://www
.nytimes.com/2009/07/10/technology/10cyber.html?_r=1; Associated Press (AP),
"U.S. Officials Eye N. Korea in Cyberattack," *USA Today,* July 9, 2009, http://
usatoday30.usatoday.com/news/washington/2009-07-08-hacking-washington
-nkorea_N.htm.

114 Officials in Seoul directly pointed their fingers at Pyongyang: Choe and Markoff, "Cyberattacks Jam Government and Commercial Web Sites in U.S. and South Korea," *New York Times,* July 9, 2009.

114 Republican lawmaker demanded: Zetter, "Lawmaker Wants 'Show of Force' Against North Korea for Website Attacks," *Wired,* July 10, 2009.

114 analysts concluded they had no evidence that North Korea or any other state was involved: Lolita C. Baldor, Associated Press (AP), "US Largely Ruling Out North Korea in 2009 Cyber Attacks," *USA Today,* July 6, 2010, http://usatoday30.usatoday .com/tech/news/computersecurity/2010-07-06-nkorea-cyber-attacks_N.htm.

114 analyst in Vietnam had earlier said that the attacks originated in the United Kingdom: Martyn Williams, "UK, Not North Korea, Source of DDOS Attacks, Researcher Says," *IDG News Service and Network World,* July 14, 2009, http://www .networkworld.com/news/2009/071409-uk-not-north-korea-source.html?ap1=rcb.

114 South Koreans insisted: "N. Korean Ministry Behind July Cyber Attacks: Spy Chief," *Yonhap News,* October 30, 2009, http://english.yonhapnews.co.kr/northkorea /2009/10/30/0401000000AEN20091030002200315.HTML.

115 semiconductors and motor vehicles to jet-propulsion technology: Michael Riley and Ashlee Vance, "Inside the Chinese Boom in Corporate Espionage," *Bloomberg BusinessWeek,* March 15, 2012, http://www.businessweek.com/articles/2012-03-14/inside -the-chinese-boom-in-corporate-espionage.

115 England's East India Company hired a Scottish botanist: "Famous Cases of Corporate Espionage," *Bloomberg BusinessWeek,* September 20, 2011, http://images.business week.com/slideshows/20110919/famous-cases-of-corporate-espionage#slide3.

116 Chinese couple in Michigan: Ed White, Associated Press (AP), "Shanshan Du, Ex-GM Worker, Allegedly Tried to Sell Hybrid Car Secrets to Chinese Companies," *Huffington Post,* July 23, 2010, http://www.huffingtonpost.com/2010/07/23 /shanshan-du-ex-gm-worker_n_656894.html.

116 Chinese employee of Valspar Corporation: "Cyber Espionage: An Economic Issue," *China Caucus* (blog), Congressional China Caucus, November 9, 2011, http:// forbes.house.gov/chinacaucus/blog/?postid=268227; *Foreign Spies Stealing U.S. Economic Secrets in Cyberspace, Report to Congress on Foreign Economic Collection and Industrial Espionage, 2009–2011,* Office of the National Counterintelligence Executive, (October 2011), 3, http://www.ncix.gov/publications/reports/fecie_all/Foreign _Economic_Collection_2011.pdf.

116 DuPont chemical researcher: "Economic Espionage," Office of the National Counterintelligence Executive, accessed October 22, 2012, http://www.ncix.gov/issues /economic/index.php.

117 "The basic premise is that when you have a network disease": Craig Mundie in discussion with the authors, November 2011.

118 In Mundie's vision: Ibid.

118 10 million lines of code: DARPA, "DARPA Increases Top Line Investment in Cyber Research by 50 Percent over next Five Years," news release, November 7, 2011, http://www.darpa.mil/NewsEvents/Releases/2011/11/07.aspx; Spencer Ackerman, "Darpa Begs Hackers: Secure Our Networks, End 'Season of Darkness,'" *Danger Room* (blog), *Wired,* November 7, 2011, http://www.wired.com/dangerroom/2011/11 /darpa-hackers-cybersecurity/.

119 "We went after the technological shifts": Regina Dugan, in discussion with the authors, July 2012.

119 They brought together cybersecurity experts: Cheryl Pellerin, American Forces

Press Service, "DARPA Goal for Cybersecurity: Change the Game," U.S. Air Force, December 20, 2010, http://www.af.mil/news/story.asp?id=123235799.

CHAPTER 4   The Future of Revolution

122   countries coming online have incredibly young populations: See low Internet penetration in 2011 for Ethiopia, Pakistan and the Philippines in "Percentage of Individuals Using the Internet," International Telecommunication Union (ITU), ICT Data and Statistics (IDS), accessed October 16, 2012, http://www.itu.int/ITU-D /ict/statistics/, and young populations for those countries as of 2011 in "Mid-Year Population by Five Year Age Groups and Sex—Custom Region—Ethiopia, Pakistan, Philippines," U.S. Census Bureau, International Data Base, accessed October 16, 2012, http://www.census.gov/population/international/data/idb/region.php.

123   women were able to play a much greater role: Courtney C. Radsch, "Unveiling the Revolutionaries: Cyberactivism and the Role of Women in the Arab Uprisings," James A. Baker III Institute for Public Policy, Rice University, May 17, 2012; Jeff Falk, "Social Media, Internet Allowed Young Arab Women to Play a Central Role in Arab Spring," May 24, 2012, Rice University, News and Media, http:// news.rice.edu/2012/05/24/social-media-and-the-internet-allowed-young-arab -women-to-play-a-central-role-in-the-arab-spring-uprisings-new-rice-study-says-2/; *Women and the Arab Spring: Taking Their Place?,* International Federation for Human Rights, accessed November 4, 2012, http://www.europarl.europa.eu /document/activities/cont/201206/20120608ATT46510/20120608ATT46510EN .pdf; Lauren Bohn, "Women and the Arab Uprisings: 8 'Agents of Change' to Follow," CNN, February 3, 2012, http://www.cnn.com/2012/02/03/world/africa/women -arab-uprisings/index.html.

123   small groups of protesters nearly every morning: Ministers in the transitional government in Tripoli in discussion with the authors, January 2012.

125   Al Jazeera English was quick to report on the number of protester deaths: "Fresh Protests Erupt in Syria," Al Jazeera, last updated April 8, 2011, http://www.aljazeera .com/news/middleeast/2011/04/201148104927711611.html.

125   Al Jazeera Arabic website did not: David Pollock, "Al Jazeera: One Organization, Two Messages," Washington Institute, Policy Analysis, April 28, 2011, http://www .washingtoninstitute.org/policy-analysis/view/aljazeera-one-organization-two -messages.

125   the disparity was due to the Arabic station's political deference to Iran: Ibid.

126   they acknowledged similar grievances: Activists from the Jasmine Revolution in discussion with the authors, January 2012.

127   Twitter account started by a twenty-something graduate student: Stephan Faris, "Meet the Man Tweeting Egypt's Voices to the World," *Time,* February 1, 2011, http://www.time.com/time/world/article/0,8599,2045489,00.html.

127   posted updates about the protests: Ibid.

127   @Jan25voices Twitter handle was a major conduit of information: Ibid.

127   Andy Carvin, who curated one of the most important streams of information: Andy Carvin, interview by Robert Siegel, "The Revolution Will Be Tweeted," NPR, February 21, 2011, http://www.npr.org/2011/02/21/133943604/The-Revolution-Will-Be -Tweeted.

130   they had formed the National Transitional Council (NTC) in Benghazi: "Anti-

Gaddafi Figures Say Form National Council," Reuters, February 27, 2011, Africa edition, http://af.reuters.com/article/idAFWEB194120110227.

130   prominent opposition figures, regime defectors, a former army official, academics, attorneys, politicians and business leaders: Dan Murphy, "The Members of Libya's National Transitional Council," *Christian Science Monitor,* September 2, 2011, http://www.csmonitor.com/World/Backchannels/2011/0902/The-members -of-Libya-s-National-Transitional-Council; David Gritten, "Key Figures in Libya's Rebel Council," BBC, August 25, 2011, http://www.bbc.co.uk/news/world-africa -12698562.

130   Citizens continued to protest the government: "Tunisia's Leaders Resign from Ruling Party," NPR, January 20, 2011, http://www.npr.org/2011/01/20/133083002 /tunisias-leaders-resign-from-ruling-party; Christopher Alexander, "Après Ben Ali: Déluge, Democracy, or Authoritarian Relapse?," *Middle East Channel* (blog), *Foreign Policy,* January 24, 2011, http://mideast.foreignpolicy.com/posts/2011/01/24 /apres_ben_ali_deluge_democracy_or_authoritarian_relapse.

130   "victim of the ministry of the interior": conversation with Tunisian prime minister Hamadi Jebali, January 2012.

130   spent fourteen years in prison: David D. Kirkpatrick, "Opposition in Tunisia Finds Chance for Rebirth," *New York Times,* January 20, 2011, http://www.nytimes .com/2011/01/21/world/africa/21islamist.html?pagewanted=all; Tarek Amara and Mariam Karouny, "Tunisia Names New Government, Scraps Secret Police," Reuters, March 8, 2011, http://in.mobile.reuters.com/article/worldNews/idINIndia -55387920110307?irpc=984.

131   "It is hard to imagine de Gaulles and Churchills appealing": Henry Kissinger in discussion with the authors, December 2011.

132   "If you are a revolutionary, show us your capabilities": Mahmoud Salem, "Chapter's End!," *Rantings of a Sandmonkey* (blog), June 18, 2012, http://www.sandmonkey .org/2012/06/18/chapters-end/.

132   He exhorted street activists to participate in governance: Mahmoud Salem, "For the Light to Come Back," *Rantings of a Sandmonkey* (blog), March 30, 2012, http://www .sandmonkey.org/2012/03/30/for-the-light-to-come-back/.

132   Tina Rosenberg's book *Join the Club*: For a more detailed interpretation of Tina Rosenberg's *Join the Club: How Peer Pressure Can Transform the World,* see Saul Austerlitz, "Power of Persuasion: Tina Rosenberg's Join the Club," review, *The National* (Abu Dhabi), February 25, 2011, http://www.thenational.ae/arts-culture /books/power-of-persuasion-tina-rosenbergs-join-the-club#full; Jeffrey D. Sachs, "Can Social Networking Cure Social Ills?," review, *New York Times,* May 20, 2011, http://www.nytimes.com/2011/05/22/books/review/book-review-join-the-club-by -tina-rosenberg.html?pagewanted=all; Thomas Hodgkinson, "Join the Club by Tina Rosenberg—Review," *Guardian* (Manchester), September 1, 2011, http://www .guardian.co.uk/books/2011/sep/02/join-club-tina-rosenberg-review; and Steve Weinberg, "C'mon, Everyone's Doing It," review, *Bookish* (blog), *Houston Chronicle,* March 27, 2011, http://blog.chron.com/bookish/2011/03/cmon-everyones-doing-it -a-review-of-tina-rosenbergs-new-book/.

132   Perhaps the most compelling evidence: Tina Rosenberg, *Join the Club: How Peer Pressure Can Transform the World* (New York: W. W. Norton and Co., 2011).

132   powerful story of Serbian activists from the past training future activists around the world: Ibid., 278–82, 332–36.

136   they subsequently *re*occupied Tahrir Square: "Egypt Anti-Military Protesters Fill

Tahrir Square," BBC, June 22, 2012, http://www.bbc.co.uk/news/world-middle
-east-18547371; Aya Batrawy, Associated Press (AP), "Egypt Protests: Thousands
Gather in Tahrir Square to Demonstrate Against Military Rule," *Huffington Post,*
April 20, 2012, http://www.huffingtonpost.com/2012/04/20/egypt-protests-tahrir
-square_n_1439802.html; Gregg Carlstrom and Evan Hill, "Scorecard: Egypt Since
the Revolution," Al Jazeera, last updated January 24, 2012, http://www.aljazeera
.com/indepth/interactive/2012/01/20121227117613598.html; "Egypt Protests: Death
Toll Up in Cairo's Tahrir Square," BBC, November 20, 2011, http://www.bbc.co.uk
/news/world-africa-15809739.

137   Iranian regime during the 2009 postelection protests: Christopher Rhoads, Geof-
frey A. Fowler, and Chip Cummins, "Iran Cracks Down on Internet Use, Foreign Media," *Wall Street Journal,* June 17, 2009, http://online.wsj.com/article
/SB124519888117821213.html.

138   Egyptian regime effectively shut down all Internet and mobile connections: James
Cowie, "Egypt Leaves the Internet," *Renesys* (blog), January 27, 2011, http://www
.renesys.com/blog/2011/01/egypt-leaves-the-internet.shtml.

138   exception to this all-ISP block: James Cowie, "Egypt Returns to the Internet," *Renesys* (blog), February 2, 2011, http://www.renesys.com/blog/2011/02/egypt-returns-to
-the-internet.shtml.

138   The country's four main Internet service providers: Cowie, "Egypt Leaves the Internet," http://www.renesys.com/blog/2011/01/egypt-leaves-the-internet.shtml.

138   mobile-phone service was also suspended: Associated Press (AP), "Vodafone: Egypt
Ordered Cell Phone Service Stopped," *Huffington Post,* January 28, 2011, http://www
.huffingtonpost.com/2011/01/28/vodafone-egypt-service-dropped_n_815493.html.

138   Vodafone Egypt, issued a statement that morning: "Statements—Vodafone Egypt,"
Vodafone, see January 28, 2011, http://www.vodafone.com/content/index/media
/press_statements/statement_on_egypt.html.

138   fiber-optic cables housed in one building in Cairo: James Glanz and John Markoff,
"Egypt Leaders Found 'Off' Switch for Internet," *New York Times,* February 15, 2011,
http://www.nytimes.com/2011/02/16/technology/16internet.html?pagewanted
=all&_r=0.

138   through its state-owned company Telecom Egypt, physically cut their service: Ibid.

138   It was a move unprecedented in recent history: Parmy Olson, "Egypt Goes Dark,
Cuts Off Internet and Mobile Networks," *Forbes,* January 28, 2011, http://www
.forbes.com/sites/parmyolson/2011/01/28/egypt-goes-dark/.

139   "Hitting one hundred percent of the population": Vittorio Colao in discussion with
the authors, August 2011.

139   "We might not have liked the request": Ibid.; see also, "Statements—Vodafone
Egypt," Vodafone, see January 28, 2011–February 3, 2011, http://www.vodafone
.com/content/index/media/press_statements/statement_on_egypt.html.

139   send out its messages over the companies' short-message-service (SMS) platform:
"Statements—Vodafone Egypt," Vodafone, see February 3, 2011, http://www
.vodafone.com/content/index/media/press_statements/statement_on_egypt.html;
Jonathan Browning, "Vodafone Says It Was Instructed to Send Pro-Mubarak
Messages to Customers," *Bloomberg,* February 3, 2011, http://www.bloomberg
.com/news/2011-02-03/vodafone-ordered-to-send-egyptian-government-messages
-update1-.html.

139   "But at a point it became incredibly political and one-sided": Vittorio Colao in discussion with the authors, August 2011.

139   "Vodafone Group PLC"—the parent company—"put out a statement": Ibid.

140   French Data Network, opened up Internet access: Jonathan Browning, "Google, Twitter Offer Egyptians Option to Tweet," *Bloomberg*, February 1, 2011, http://www .bloomberg.com/news/2011-01-31/egyptians-turn-to-dial-up-service-to-get-around -government-s-web-shutdown.html.

140   Google launched a tweet-by-phone service: Ujjwal Singh and AbdelKarim Mardini, "Some Weekend Work That Will (Hopefully) Enable More Egyptians to Be Heard," *Google Blog,* January 31, 2011, http://googleblog.blogspot.com/2011/01/some -weekend-work-that-will-hopefully.html.

140   "We decided that this has to be discussed": Vittorio Colao in discussion with the authors, August 2011.

142   Egyptian police's vice squad would troll chat rooms: *In a Time of Torture: The Assault on Justice in Egypt's Crackdown on Homosexual Conduct,* Human Rights Watch (HRW): 2004, http://www.hrw.org/en/reports/2004/02/29/time-torture.

142   Cairo vice squad raided a floating nightclub: Ibid.; "Egypt: Egyptian Justice on Trial—The Case of the Cairo 52," International Gay and Lesbian Human Rights Commission, October 15, 2001, http://www.iglhrc.org/cgi-bin/iowa/article/take action/partners/692.html.

142   a Chinese version of the protests: Andrew Jacobs, "Chinese Government Responds to Call for Protests," *New York Times,* February 20, 2011, http://www.nytimes.com /2011/02/21/world/asia/21china.html?_r=1.

143   retaliated against a group of women: "Rights Group Decries Flogging Sentence for Female Saudi Driver," CNN, September 27, 2011, http://articles.cnn.com /2011-09-27/middleeast/world_meast_saudi-arabia-flogging_1_flogging-sentence -women2drive-saudi-woman?_s=PM:MIDDLEEAST.

143   As news of her sentence spread: Ibid.; Amnesty International (AI), "Flogging Sentence for Saudi Arabian Woman After Driving 'Beggars Belief,'" press release, September 27, 2011, https://www.amnesty.org/en/for-media/press-releases/flogging -sentence-saudi-arabian-woman-after-driving-%E2%80%9Cbeggars-belief %E2%80%9D-2011-0.

143   led the government to revoke the decision: "Saudi King Revokes Flogging of Female Driver," CNN, September 29, 2011, http://www.cnn.com/2011/09/28/world/meast /saudi-arabia-flogging/index.html.

144   a decision to ban a satirical short film: Prince Alwaleed bin Talal al-Saud in discussion with the authors, February 2011; Faisal J. Abbas, "Monopoly: The Saudi Short-Film Which Went a Long Way," *Huffington Post,* September 9, 2011, http://www .huffingtonpost.com/faisal-abbas/monopoly-the-saudi-shortf_b_969540.html.

144   The film, *Monopoly,* appeared on YouTube: Prince Alwaleed bin Talal al-Saud in discussion with the authors, February 2011. The film appeared on both Facebook and YouTube. The prince discussed the Facebook appearance of the video.

144   accumulated more than a million views: Ibid.

144   one of the highest rates of YouTube playbacks: "Saudi Arabia Ranks First in You-Tube Views," Al Arabiya, May 22, 2012, http://english.alarabiya.net/articles/2012 /05/22/215774.html; Simon Owens, "Saudi Satire Ignites YouTube's Massive Growth in Middle East," *U.S. News,* May 30, 2012, http://www.usnews.com/news/articles /2012/05/30/saudi-satire-ignites-youtubes-massive-growth-in-middle-east.

144   fastest growing mobile market anywhere: *African Mobile Observatory 2011: Driving Economic and Social Development Through Mobile Services,* Groupe Speciale Mobile

(GSM), 9, accessed October 17, 2011, http://www.gsma.com/publicpolicy/wp
-content/uploads/2012/04/africamobileobservatory2011-1.pdf.

146 "The Internet is good for letting off steam": Prime Minister Lee Hsien Loong in
discussion with the authors, November 2011.

146 Young people everywhere: Prime Minister Lee Hsien Loong's assertion that young
people want to be cool is supported by Tina Rosenberg's discussion of the need to
be cool as a key part of Otpor's strategy, which it has taught to opposition groups
around the world. For examples of the "coolness factor" in Otpor, see Rosenberg,
Join the Club, 223–224, 229, 256–58, 260, 276.

146 "Currygate": Shamim Adam, "Singapore Curry Protest Heats Up Vote with Facebook Campaign," *Bloomberg*, August 19, 2011, http://www.bloomberg.com/news
/2011-08-18/singapore-curry-protest-heats-up-vote.html; "Singaporeans to Launch
Largest 'Protest' over 'Currygate' Incident," *TR Emeritus* (blog), August 21, 2011,
http://www.tremeritus.com/2011/08/21/singaporeans-to-launch-largest-protest-over
-currygate-incident/.

146 "A Chinese immigrant and a Singaporean of Indian descent quarreled": Lee Hsien
Loong in discussion with the authors, November 2011.

147 almost a billion Chinese citizens: Michael Kan, International Data Group (IDG)
News Service, "China's Internet Population Reaches 538 Million," July 19, 2012,
PCWorld, http://www.pcworld.com/article/259482/chinas_internet_population
_reaches_538_million.html; at the time of writing, China's population exceeded
1.3 billion, so there were approximately 800 million Chinese citizens left to become
connected. We factored in population increase projections over the next decade to
estimate almost a billion. According to a 2012 report from the Committee to Protect
Journalists, Eritrea was most censored, followed by North Korea.

147 "What happens in China is beyond anyone's full control": Lee Hsien Loong in discussion with the authors, November 2011.

148 "The history of revolutions is a confluence": Henry Kissinger in discussion with the
authors, December 2011.

CHAPTER 5   The Future of Terrorism

151 we're acutely vulnerable to cyber terrorism: There is some overlap in tactics between
cyber terrorism and criminal hacking, but generally the motivations distinguish the
two. This is not unlike the distinction that is made between narco-trafficking and
terrorism.

152 one of the greatest shared fears among American troops: Army captain (in Iraq) in
discussion with the authors, November 2009.

152 The IED of 2009 was cheaper and more innovative: Ibid.

152 What was once a sophisticated and lucrative violent activity: One of the authors first
learned about this while speaking on a panel with Jonathan Powers at Johns Hopkins
School of Advanced International Studies in 2005. The authors have since corroborated this data point with additional anecdotes from civilian and military officials
who have been working on or deployed in Iraq over the past decade.

153 "maker phenomenon": Andy Rubin in discussion with the authors, February 2012.

157 Somalia's al-Shabaab insurgent group on Twitter: Will Oremus, "Twitter of Terror,"
Slate, December 23, 2011, http://www.slate.com/articles/technology/technocracy

/2011/12/al_shabaab_twitter_a_somali_militant_group_unveils_a_new_social
_media_strategy_for_terrorists_.html.

157 Anwar al-Awlaki, the late American-born extremist cleric: "Profile: Anwar al-Awlaki," Anti-Defamation League (ADL), updated November 2011, http://www.adl.org/main_Terrorism/anwar_al-awlaki.htm.

157 several successful and would-be terrorists cited him: Pierre Thomas, Martha Raddatz, Rhonda Schwartz and Jason Ryan, "Fort Hood Suspect Yells Nidal Hasan's Name in Court," *ABC Blotter,* July 29, 2011, http://abcnews.go.com/Blotter/fort-hood-suspect-naser-jason-abdo-yells-nidal-hasan/story?id=14187568#.UIIwW8VG-8C; Bruce Hoffman, "Why al Qaeda Will Survive," *Daily Beast,* September 30, 2011, http://www.thedailybeast.com/articles/2011/09/30/al-awlaki-s-death-nothing-more-than-a-glancing-blow-al-qaeda-stronger-than-everest.html.

157 "Even the most anti-Western religious figures in Saudi Arabia": Prince Alwaleed bin Talal al-Saud in discussion with the authors, February 2012.

157 "We pitched propaganda stalls outside the Motorola offices": Maajid Nawaz in discussion with the authors, February 2012.

158 Colombian prison officials stopped an eleven-year-old girl: "Colombia Catches Girl 'Smuggling 74 Mobiles into Jail,'" BBC, February 6, 2011, http://www.bbc.co.uk/news/world-latin-america-12378390.

158 In Brazil, inmates trained carrier pigeons: "Pigeons Fly Mobile Phones to Brazilian Prisoners," *Telegraph* (London), March 30, 2009, http://www.telegraph.co.uk/news/newstopics/howaboutthat/5079580/Pigeons-fly-mobile-phones-to-Brazilian-prisoners.html.

158 local gang hired a teenager: Associated Press (AP), "Police: Brazilian Teen Used Bow and Arrow to Launch Illegal Cell Phones over Prison Walls," *Fox News,* September 2, 2010, http://www.foxnews.com/world/2010/09/02/police-brazilian-teen-used-bow-arrow-launch-illegal-cell-phones-prison-walls/.

158 going rate for a contraband smart phone: Former member of a South Central Los Angeles gang in discussion with the authors, April 2012.

159 Afghanistan, a country with one of the lowest rates of connectivity in the world: "Mobile-Cellular Subscriptions" and "Percentage of Individuals Using the Internet," International Telecommunication Union (ITU), ICT Data and Statistics (IDS), accessed October 19, 2012, http://www.itu.int/ITU-D/ict/statistics/.

159 tens of thousands of political prisoners were killed there annually: The authors received this information during an unclassified briefing with prison staff, February 2009.

159 terrorist nerve center: Rod Nordland and Sharifullah Sahak, "Afghan Government Says Prisoner Directed Attacks," *New York Times,* February 10, 2011, http://www.nytimes.com/2011/02/11/world/asia/11afghan.html?_r=1&scp=3&sq=pul%20e%20charki&st=cse.

159 Following a violent riot in 2008 in the prison's Cell Block Three: Description of the terror cell operating from Pul-e-Charkhi comes from Jared's briefings (unclassified) and interviews during his visit to the prison in February 2009; see also Joshua Philipp, "Corruption Turning Afghan Prisons into Taliban Bases: Imprisoned Taliban Leaders Coordinate Attacks from Within Prison Walls," *Epoch Times,* August 29, 2011, http://www.theepochtimes.com/n2/world/corruption-turning-afghan-prisons-into-taliban-bases-60910.html.

160 Agie responded to a joking request for his phone number: Mullah Akbar Agie in discussion with Jared Cohen, February 2009.

162   Shortly thereafter, a series of cyber attacks crippled: "Anonymous (Internet Group),"
New York Times, updated March 8, 2012, http://topics.nytimes.com/top/reference/
timestopics/organizations/a/anonymous_internet_group/index.html.

162   vowed to take revenge on any organization: Sean-Paul Correll, "Operation: Pay-
back Broadens to Operation Avenge Assange." *Pandalabs* (blog), December 6, 2010,
http://pandalabs.pandasecurity.com/operationpayback-broadens-to-operation
-avenge-assange/; Mathew Ingram, "WikiLeaks Gets Its Own 'Axis of Evil' Defense
Network," *GigaOM* (blog), December 8, 2010, http://gigaom.com/2010/12/08
/wikileaks-gets-its-own-axis-of-evil-defence-network/.

163   A string of global investigations followed: U.S. Department of Justice, "Sixteen
Individuals Arrested in the United States for Alleged Roles in Cyber Attacks,"
national press release, July 19, 2011, http://www.fbi.gov/news/pressrel/press-releases
/sixteen-individuals-arrested-in-the-united-states-for-alleged-roles-in-cyber-attacks;
Andy Greenberg, "Fourteen Anonymous Hackers Arrested for 'Operation Avenge
Assange,' LulzSec Leader Claims He's Not Affected," *Forbes,* July 19, 2011, http://
www.forbes.com/sites/andygreenberg/2011/07/19/anonymous-arrests-continue
-lulzsec-leader-claims-hes-not-affected/; "Hackers Arrested in US, NL and UK,"
Radio Netherlands Worldwide, July 20, 2011, http://www.rnw.nl/english/bulletin
/hackers-arrested-us-nl-and-uk.

163   he told *The New York Times* via e-mail: Somini Sengupta, "Hacker Rattles Security
Circles," *New York Times,* September 11, 2011, http://www.nytimes.com/2011/09/12
/technology/hacker-rattles-internet-security-circles.html?pagewanted=all&_r=0.

164   Boasting aside, Comodohacker was able to forge: Ibid.

164   compromised the communications: Ibid.

164   He said he attacked: Ibid.

164   declared that he was "one of the strongest haters of Israel": "I Will Finish Israel Off
Electronically: Ox-Omar," *Emirates 24/7,* January 22, 2012, http://www.emirates247
.com/news/world/i-will-finish-israel-off-electronically-ox-omar-2012-01-22-1
.438856.

164   a file that contained four hundred thousand credit-card numbers: Chloe Albanesius,
"Hackers Target Israeli Stock Exchange, Airline Web Sites," *PC Magazine,* January 16, 2012, http://www.pcmag.com/article2/0,2817,2398941,00.asp.

164   most of these were duplicates: Isabel Kershner, "Cyberattack Exposes 20,000 Israeli
Credit Card Numbers and Details About Users," *New York Times,* January 6,
2012, http://www.nytimes.com/2012/01/07/world/middleeast/cyberattack-exposes
-20000-israeli-credit-card-numbers.html.

164   He claimed to represent a group of Wahhabi hackers: Jonathon Blakeley, "Israeli
Credit Card Hack," *deLiberation,* January 5, 2012, http://www.deliberation.info
/israeli-credit-card-hack/.

164   "It will be so fun to see": Ehud Kenan, "Saudi Hackers Leak Personal Information
of Thousands of Israelis," *YNet,* January 3, 2012, http://www.ynetnews.com/articles
/0,7340,L-4170465,00.html.

164   the websites of Israel's El Al Airlines and its stock exchange were brought down:
Isabel Kershner, "2 Israeli Web Sites Crippled as Cyberwar Escalates," *New York
Times,* January 16, 2012, http://www.nytimes.com/2012/01/17/world/middleeast
/cyber-attacks-temporarily-cripple-2-israeli-web-sites.html.

164   attacks would be reduced if Israel apologized for its "genocide": Yaakov Lappin, "'I
Want to Harm Israel,' Saudi Hacker Tells 'Post,'" *Jerusalem Post,* January 16, 2012, http://
www.jpost.com/NationalNews/Article.aspx?id=253893; Saar Haas, "'OxOmar'

Demands Israeli Apology," *YNet,* January 16, 2012, http://www.ynetnews.com /articles/0,7340,L-4176436,00.html?utm_source=dlvr.it&utm_medium=twitter.

164 "badge of honor that I have been personally targeted": Danny Ayalon's Facebook page, posts on January 13 and 16, 2012, accessed October 20, 2012, https://www .facebook.com/DannyAyalon.

168 DARPA approved eight contracts: Austin Wright, "With Cyber Fast Track, Pentagon Funds Hacker Research," *Politico,* December 7, 2011, http://www.politico.com /news/stories/1211/70016.html.

168 "democratized, crowd-sourced innovation": Statement by Dr. Regina E. Dugan, submitted to the Subcommittee on Terrorism, Unconventional Threats and Capabilities of the House Armed Services Committee, United States House of Representatives, March 23, 2010, www.darpa.mil/WorkArea/DownloadAsset.aspx?id=542.

168 "There is a sense among many that hackers and Anonymous are just evildoers": Regina Dugan, in discussion with the authors, July 2012.

169 lack of Internet access in a large urban home: Mark Mazzetti and Helene Cooper, "Detective Work on Courier Led to Breakthrough on bin Laden," *New York Times,* May 2, 2011, http://www.nytimes.com/2011/05/02/world/asia/02reconstruct -capture-osama-bin-laden.html; Bob Woodward, "Death of Osama bin Laden: Phone Call Pointed U.S. to Compound—and to 'The Pacer,'" *Washington Post,* May 6, 2011, http://www.washingtonpost.com/world/national-security/death-of -osama-bin-laden-phone-call-pointed-us-to-compound—and-to-the-pacer/2011 /05/06/AFnSVaCG_story.html.

169 But when Navy SEAL Team Six raided his home: Joby Warrick, "Al-Qaeda Data Yield Details of Planned Plots," *Washington Post,* May 5, 2011, http:// www.washingtonpost.com/world/national-security/al-qaeda-data-yields -details-of-planned-plots/2011/05/05/AFFQ3L2F_story.html.; Woodward, "Death of Osama bin Laden," http://www.washingtonpost.com/world/national-security/death -of-osama-bin-laden-phone-call-pointed-us-to-compound—and-to-the-pacer/2011 /05/06/AFnSVaCG_story.html.

169 Mumbai attacks: Hari Kumar, "India Says Pakistan Aided Planner of Mumbai Attacks," *New York Times,* June 27, 2012, http://www.nytimes.com/2012/06 /28/world/asia/india-says-pakistan-aided-abu-jindal-in-mumbai-attacks.html; Harmeet Shah Singh, "India Makes Key Arrest in Mumbai Terror Plot," CNN, June 26, 2012, http://articles.cnn.com/2012-06-26/asia/world_asia_india-terror -arrest_1_fahim-ansari-ujjwal-nikam-sabauddin-ahmed?_s=PM:ASIA; "Mumbai Attacks 'Handler' Arrested in India," Agence France-Presse (AFP), June 25, 2012, http://www.google.com/hostednews/afp/article/ALeqM5gydBxOITFOjQ _gOjs278EF2DTvIQ?docId=CNG.1ec8f11cdfb59279e03f13dafbcd927a.01.

To better understand the role that technology played in the 2008 Mumbai attacks, we spoke to Prakash V. Shukla, who is a senior vice president and the chief information officer for Taj Hotels Resorts and Palaces, which operates the Taj Mahal Hotel. He explained, "It was very clear from reviewing CCTV footage these individuals had never been to the Taj. However, they knew exactly where things were in the hotel, they knew how to get around, etc. The old part of the hotel was built over [a] hundred years ago, therefore we didn't have floor plans. Combination of the Taj website, Google Maps gave one a fairly good idea of the layout of the hotel. The website also describes the location of premium rooms, which are located on upper floors. It was very easy for them to plan attacks of high-profile targets like the Taj,

Oberoi, the train station, etc. This, augmented with Hedley's reports from actual reconnaissance work in India, gave the terrorists a fairly good idea of the locations. From the time attacks started, the terrorists immediately moved to the old wing (premium) of the hotel and started moving towards the upper floors. Satellite radios were procured. Along with several money transfers using electronic fund distribution. In India, several prepaid SIM cards were procured."

High-profile terrorist attacks on commercial hotels will have implications for how the hospitality industry deals with security. In speaking with Shukla, we learned that "the hotel industry is going the way of the airline industry. The same measures that the airlines have taken in terms of scanning baggage, doing reference checks on passengers, are being carried out in the hotel industry. Taj, specifically, has a team from Israel who has consulted with us for over four years in creating a security architecture to prevent these types of attacks. In 2008, we had security but the guards were unarmed. We learned that the police were woefully inadequate to handle the situation, and by the time the NSG teams and the Marcos teams arrived it was already over twelve hours. The overall security architecture has several components. Profiling: Guests are profiled from incoming-arrivals list and security agencies are notified of the arrivals; walking in, guests are profiled, and each of our security folks [is] trained to observe individuals entering the hotel; all baggage is screened; disaster-planning drills are conducted on regular intervals; staff has been trained to observe; we now have armed security personnel who are in civilian clothes; all hotel security staff has gone through a month training in Israel on handling of firearms and handling of situations. We have spent [a] significant amount of money to put these measures in place, and we feel that our target would be harder than [the] rest of the hotel industry, and therefore we feel fairly confident that our hotels may not be retargeted. However, having said that, this is a dynamic situation. Whilst we grow in sophistication so do our enemies, and we have to constantly innovate and keep improving the security."

169　The gunmen relied on basic consumer technologies: Jeremy Kahn, "Mumbai Terrorists Relied on New Technology for Attacks," *New York Times,* December 8, 2008, http://www.nytimes.com/2008/12/09/world/asia/09mumbai.html; Damien McElroy, "Mumbai Attacks: Terrorists Monitored British Websites Using BlackBerry Phones," *Telegraph* (London), November 28, 2008, http://www.telegraph .co.uk/news/worldnews/asia/india/3534599/Mumbai-attacks-Terrorists-monitored -coverage-on-UK-websites-using-BlackBerry-phones-bombay-india.html.

170　electronic trail: "Global Lessons from the Mumbai Terror Attacks," Investigative Project on Terrorism (IPT), November 25, 2009, http://www.investigativeproject .org/1539/global-lessons-from-the-mumbai-terror-attacks.

171　described a top al-Qaeda commander who was exceptionally cautious: A Navy SEAL Team Six member in discussion with the authors, February 2012.

172　Canadian journalist Amanda Lindhout was kidnapped: "Canadian Amanda Lindhout Freed in Somalia," CBC (Ottawa), last updated November 25, 2009, http:// www.cbc.ca/news/world/story/2009/11/25/amanda-lindhout-free.html.

172　her former captors: Author conversation with Amanda Lindhout, July 2012.

172　It is estimated that more than 90 percent of people worldwide: *Technology / Internet Trends* October 18, 2007, Morgan Stanley (China Mobile 50K Survey), 7. Posted on Scribd, http://www.scribd.com/doc/404905/Mary-Meeker-Explains-The-Internet.

172　Osama bin Laden's compound contained a large stash of pornographic videos: Scott

Shane, "Pornography Is Found in bin Laden Compound Files, U.S. Officials Say," *New York Times,* May 13, 2011, http://www.nytimes.com/2011/05/14/world/asia /14binladen.html.

174 Secretariat of Public Security compound in Mexico City: Venu Sarakki et al., "Mexico's National Command and Control Center Challenges and Successes," 16th International Command and Control Research and Technology Symposium in Quebec, Canada, June 21–23, 2011, http://www.dtic.mil/dtic/tr/fulltext/u2/a547202.pdf.

175 TIA was designed and funded to aggregate all "transactional" data: Dr. John Poindexter, "Overview of the Information Awareness Office." Remarks as prepared for the DARPATech 2002 Conference, August 2, 2002. Posted by the Federation of American Scientists (FAS), http://www.fas.org/irp/agency/dod/poindexter.html.

175 provision to deny all funds for the program: Department of Defense Appropriations Act, 2004, S.1382, 108th Cong. (2003), see Sec. 8120; Department of Defense Appropriations Act, 2004, H.R.2658, 108th Cong. (2003) (Enrolled Bill), see Sec. 8131.

175 some of its projects later found shelter: Associated Press (AP), "U.S. Still Mining Terror Data," *Wired,* February 23, 2004, http://www.wired.com/politics/law/news /2004/02/62390; Michael Hirsh, "Wanted: Competent Big Brothers," *Newsweek and Daily Beast,* February 8, 2006, http://www.thedailybeast.com/newsweek/2006 /02/08/wanted-competent-big-brothers.html.

178 An estimated 52 percent of the world's population is under the age of thirty: "Mid-Year Population by Five Year Age Groups and Sex—World, 2011," U.S. Census Bureau, International Data Base, accessed October 20, 2012, http://www.census .gov/population/international/data/idb/informationGateway.php.

179 "What defeats terrorism is really two things": General Stanley McChrystal, interview by Susanne Koelbl, "Killing the Enemy Is Not the Best Route to Success," *Der Spiegel,* January 11, 2010, http://www.spiegel.de/international/world/spiegel-interview -with-general-stanley-mcchrystal-killing-the-enemy-is-not-the-best-route-to -success-a-671267.html.

181 With more than four billion videos viewed daily: Alexei Oreskovic, "Exclusive: You-Tube Hits 4 Billion Daily Video Views," Reuters, January 23, 2012, http://www .reuters.com/article/2012/01/23/us-google-youtube-idUSTRE80M0TS20120123.

CHAPTER 6 The Future of Conflict, Combat and Intervention

183 "orient [us] away from violence and toward cooperation and altruism": Steven Pinker, *The Better Angels of Our Nature: Why Violence Has Declined* (New York: Viking, 2011), xxv.

183 "The world begins to look different": Ibid., xxvi.

186 deliberately excludes some 2.2 million ethnic Roma: Amnesty International (AI), "Romania Must End Forced Evictions of Roma Families," press release, January 26, 2010, http://www.amnesty.org/en/for-media/press-releases/romania-must-end -forced-evictions-roma-families-20100126.

The Roma are persecuted in similar fashion across Eastern Europe and, increasingly, Western Europe. In July 2010, President Nicolas Sarkozy of France led a campaign to forcibly repatriate his country's foreign Roma population to Bulgaria and Romania. Within a month, more than fifty illegal Romany camps had been closed, and by September over a thousand Roma had been deported. See "France Sends Roma Gypsies Back to Romania," BBC, August 20, 2010, http://www.bbc.co.uk

/news/world-europe-11020429; "France: Renewed Crackdown on Roma: End Discriminatory Roma Camp Evictions and Removals," Human Rights Watch (HRW), News, August 10, 2010, http://www.hrw.org/news/2012/08/10/france-renewed -crackdown-roma; "French Ministers Fume After Reding Rebuke Over Roma," BBC, September 15, 2010, http://www.bbc.co.uk/news/world-europe-11310560.

188   "Online intimidation by hate groups": Christian Picciolini in discussion with the authors, April 2012.

189   Julius Caesar: Julius Caesar, *The Gallic Wars,* translation by John Warrington with a preface by John Mason Brown and an introduction by the translator (Norwalk, Conn.: Easton Press, 1983); see also Dr. Neil Faulkner, "The Official Truth: Propaganda in the Roman Empire," BBC, History, last updated February 17, 2011, http:// www.bbc.co.uk/history/ancient/romans/romanpropaganda_article_01.shtml.

189   "Video: IDF pilots wait for area to be clear": @IDFspokesperson tweet, November 19, 2012.

192   Neda Agha-Soltan: Nazila Fathi, "In a Death Seen Around the World, a Symbol of Iranian Protests," *New York Times,* June 22, 2009, http://www.nytimes.com/2009 /06/23/world/middleeast/23neda.html.

192   The videos were passed: Thomas Erdbrink, "In Iran, a Woman Named Neda Becomes Opposition Icon in Death," *Washington Post,* June 23, 2009, http://www .washingtonpost.com/wp-dyn/content/article/2009/06/22/AR2009062203041 .html.

193   raided their offices: Information obtained in Jared Cohen's research for his book *One Hundred Days of Silence: America and the Rwanda Genocide* (Lanham: Rowman & Littlefield Publishers, 2007); see also Alison Liebhafsky Des Forges, *Leave None to Tell the Story: Genocide in Rwanda* (New York: Human Rights Watch, 1999).

193   Hutu radio stations announced names and addresses: Allan Thompson, ed., with a statement by Kofi Annan, *The Media and the Rwanda Genocide* (London: Pluto Press, 2007), 49, http://www.internews.org/sites/default/files/resources/TheMedia&The RwandaGenocide.pdf.

196   "virtually inoperable": Dan Verton, "Serbs Launch Cyberattack on NATO," *Federal Computer Week,* April 4, 1999, http://fcw.com/articles/1999/04/04/serbs-launch -cyberattack-on-nato.aspx.

197   Tom Downey's revealing March 2010 article: Tom Downey, "China's Cyberposse," *New York Times Magazine,* March 3, 2010, http://www.nytimes.com/2010/03/07 /magazine/07Human-t.html.

197   gruesome video: Ibid.

197   perpetrator was soon tracked: Ibid.

198   took just six days: Ibid.

198   a bill (struck down one month later by the French Constitutional Council): Scott Sayare, "French Council Strikes Down Bill on Armenian Genocide Denial," *New York Times,* February 28, 2012, http://www.nytimes.com/2012/02/29/world /europe/french-bill-on-armenian-genocide-is-struck-down.html.

198   "racist and discriminatory": "Turkey PM Says French Bill on Genocide Denial 'Racist,'" BBC, January 24, 2012, http://www.bbc.co.uk/news/world-europe-16695133.

201   "singularity": P. W. Singer, *Wired for War: The Robotics Revolution and Conflict in the 21st Century* (New York: Penguin Press, 2009), 102.

201   DARPA's mission: DARPA, "About," accessed October 9, 2012, http://www.darpa .mil/About.aspx; DARPA, "Our Work," accessed October 9, 2012, http://www .darpa.mil/our_work/.

202 three Ds: Singer, *Wired for War*, 63.

202 iRobot, the company that invented: Ibid., 21–23.

202 Two PackBots were deployed during the Fukushima nuclear crisis: Amar Toor, "iRobot Packbots Enter Fukushima Nuclear Plant to Gather Data, Take Photos, Save Lives," *Engadget*, April 18, 2011, http://www.engadget.com/2011/04/18/irobot -packbots-enter-fukushima-nuclear-plant-to-gather-data-ta/.

202 Foster-Miller, makes a PackBot competitor: Singer, *Wired for War*, 26.

202 And then there are the aerial drones: For descriptions of the Predator, Raven and Reaper drones, see Singer, *Wired for War*, 32–35, 37, 116.

203 31 percent of all military aircraft: Spencer Ackerman and Noah Shachtman, "Almost 1 in 3 U.S. Warplanes Is a Robot," *Danger Room* (blog), *Wired*, January 9, 2012, http://www.wired.com/dangerroom/2012/01/drone-report/.

203 "lethal kinetics"—operations involving fire—"will be handed over to bots": Harry Wingo, in discussion with the authors, April 2012.

203 SWORDS robots: Singer, *Wired for War*, 29–32; Noah Shachtman, "First Armed Robots on Patrol in Iraq (Updated)," *Danger Room* (blog), *Wired*, August 2, 2007, http://www.wired.com/dangerroom/2007/08/httpwwwnational/.

203 combat units in the future: Navy SEAL in discussion with the authors, February 2012.

204 "It's a big strategic question for them": Peter Warren Singer in discussion with the authors, April 2012.

205 Joint Tactical Radio System: Bob Brewin, "Pentagon Shutters Joint Tactical Radio System Program Office," *Nextgov*, August 1, 2012, http://www.nextgov.com/mobile/ 2012/08/pentagon-shutters-joint-tactical-radio-system-program-office/57173/; Matthew Potter, Defense Procurement News, "Joint Program Executive Office Joint Tactical Radio System (JPEO JTRS) Stands Down and Joint Tactical Networking Center (JTNC) Opens," press release, October 1, 2012, http://www.defenseprocurement news.com/2012/10/01/joint-program-executive-office-joint-tactical-radio-system -jpeo-jtrs-stands-down-and-joint-tactical-networking-center-jtnc-opens-press -release/.

205 "They just can't afford that kind of process anymore": Peter Warren Singer in discussion with the authors, April 2012.

205 "The military was, in some ways": Ibid.

206 Even Venezuela has joined the club: Brian Ellsworth, "Venezuela Says Building Drones with Iran's Help," Reuters, June 14, 2012, http://www.reuters.com/article /2012/06/14/us-venezuela-iran-drone-idUSBRE85D14N20120614.

206 "Of course we're doing it": Robert Beckhusen, "Iranian Missile Engineer Oversees Chavez's Drones," *Danger Room* (blog), *Wired*, June 18, 2012, http://www.wired .com/dangerroom/2012/06/mystery-cargo/.

206 "Most advances in technology, particularly big ones, tend to make people nervous": Regina Dugan, in discussion with the authors, July 2012.

207 "non-state actors that range": Peter Warren Singer in discussion with the authors, April 2012.

207 unmanned drones, available to rent for surveillance: Singer, *Wired for War*, 265.

207 In 2009, it was contracted to load bombs: James Risen and Mark Mazzetti, "C.I.A. Said to Use Outsiders to Put Bombs on Drones," *New York Times*, August 20, 2009, http://www.nytimes.com/2009/08/21/us/21intel.html.

207 For example, some real-estate firms are now using private drones: Somini Sengupta,

"Who Is Flying Drones over America?," *Bits* (blog), *New York Times,* July 14, 2012, http://bits.blogs.nytimes.com/2012/07/14/who-is-flying-drones-over-america/.

207 Kansas State University has established a degree: Jefferson Morley, "Drones Invade Campus," *Salon,* May 1, 2012, http://www.salon.com/2012/05/01/drones_on _campus/.

208 "battle of persuasion": Peter Warren Singer, quoted by Noah Shachtman, "Insurgents Intercept Drone Video in King-Size Security," *Danger Room* (blog), *Wired,* December 17, 2009, http://www.wired.com/dangerroom/2009/12/insurgents-intercept -drone-video-in-king-sized-security-breach/.

209 RQ-170 Sentinel: Scott Peterson, "Downed U.S. Drone: How Iran Caught the 'Beast,'" *Christian Science Monitor,* December 9, 2011, http://www.csmonitor.com /World/Middle-East/2011/1209/Downed-US-drone-How-Iran-caught-the-beast.

209 "land on its own where we wanted it to": Scott Peterson and Payam Faramarzi, "Exclusive: Iran Hijacked U.S. Drone, Says Iranian Engineer," *Christian Science Monitor,* December 15, 2011, http://www.csmonitor.com/World/Middle-East/2011 /1215/Exclusive-Iran-hijacked-US-drone-says-Iranian-engineer-Video.

209 known as spoofing: Adam Rawnsley, "Iran's Alleged Drone Hack: Tough, but Possible," *Danger Room* (blog), *Wired,* December 16, 2011, http://www.wired.com /dangerroom/2011/12/iran-drone-hack-gps/.

209 $6 million: Dan Murphy, "Obama Taking Heat for Asking for U.S. Drone Back? Pay Little Heed," *Christian Science Monitor,* December 15, 2011, http://www.csmonitor .com/World/Backchannels/2011/1215/Obama-taking-heat-for-asking-for-US-drone -back-Pay-little-heed.

210 leaks resulted in detailed articles: Daniel Klaidman, "Drones: How Obama Learned to Kill," May 28, 2012, *Newsweek and Daily Beast,* http://www.thedailybeast.com /newsweek/2012/05/27/drones-the-silent-killers.html; Jo Becker and Scott Shane, "Secret 'Kill List' Proves a Test of Obama's Principles and Will," *New York Times,* May 29, 2012, http://www.nytimes.com/2012/05/29/world/obamas-leadership-in -war-on-al-qaeda.html; David E. Sanger, "Obama Order Sped Up Wave of Cyberattacks Against Iran," *New York Times,* June 1, 2012, http://www.nytimes.com/2012 /06/01/world/middleeast/obama-ordered-wave-of-cyberattacks-against-iran.html; Charlie Savage, "Holder Directs U.S. Attorneys to Track Down Paths of Leaks," *New York Times,* June 8, 2012, http://www.nytimes.com/2012/06/09/us/politics /holder-directs-us-attorneys-to-investigate-leaks.html?pagewanted=all.

212 the Bosnian campaign of the 1990s: Siobhan Gorman, Yochi J. Dreazen and August Cole, "Insurgents Hack U.S. Drones," *Wall Street Journal,* December 17, 2009, http://online.wsj.com/article/SB126102247889095011.html.

212 U.S. military troops first discovered laptops: Ibid.

212 As Peter Singer pointed out, during World War I, when the tank first appeared: Peter Warren Singer in discussion with the authors, April 2012.

212 targeted Soviet tanks: Abdul Rahim Wardak in discussion with the authors, June 2012.

213 "The ground robots our soldiers use": Peter Warren Singer in discussion with the authors, April 2012.

213 Faced with this decision: Ibid.

214 Responsibility to Protect (RtoP): Jayshree Bajoria, "Libya and the Responsibility to Protect," Counsel on Foreign Relations, analysis brief, March 24, 2011, http://www .cfr.org/libya/libya-responsibility-protect/p24480.

215   a brave American soul called Fred: Libyan ministers in discussion with the authors, June 2012.

215   Bangladesh is among the most frequent contributors of troops: "Ranking of Military and Police Contributions to U.N. Operations," United Nations Peacekeeping, Resources, August 31, 2012, http://www.un.org/en/peacekeeping/contributors/2012 /august12_2.pdf.

CHAPTER 7   The Future of Reconstruction

218   a small GSM . . . network limited to government officials: "Apple's iPhone and Afghanistan's Taliban," *Cellular-News* (London), February 13, 2009, http://www .cellular-news.com/story/36027.php.

219   Saddam Hussein banned mobile phones entirely: W. David Gardner, "For Sale: Iraq's Cell-Phone Franchises," *InformationWeek,* July 27, 2005, http://www.information week.com/news/166403218.

219   combatants in the ensuing conflicts were the only ones: Author discussions with members of the Libyan ministry of communications and informatics, June 2012.

219   MTC-Vodafone, a regional telecom company: "Post-War Telecommunications Developments in Iraq," Office of Technology and Electronic Commerce, Research by Country/Region, accessed October 18, 2012, http://web.ita.doc.gov/ITI/itiHome .nsf/6502bd9adeb499b285256cdb00685f77/e781b255ae7a4f9a85256d9c0068abd9 ?OpenDocument.

219   MCI, got the nod in Baghdad: Ibid.

219   towers were put up all over the country: Senior CPA (Coalition Provisional Authority) official in discussion with the authors, January 2011.

219   the sector was booming: *Iraq—Telecoms, Mobile, Broadband and Forecasts: Executive Summary,* BuddeComm, accessed October 18, 2012, http://www.budde.com.au /Research/Iraq-Telecoms-Mobile-Broadband-and-Forecasts.html.

219   U.N. established a mobile network: "Press Briefing by the U.N. Offices for Pakistan and Afghanistan," United Nations, News Centre, January 16, 2001, http://www.un .org/apps/news/infocus/afghanistan/infocusnews.asp?NewsID=136&sID=4.

219   four major operators in Afghanistan: *Afghanistan—Telecoms, Mobile, Internet and Forecasts: Executive Summary,* BuddeComm, accessed October 18, 2012. http://www .budde.com.au/Research/Afghanistan-Telecoms-Mobile-Internet-and-Forecasts .html. The above source shows that as of 2011 there were 17.6 million total mobile subscribers in Afghanistan with the four major operators "carrying market shares in excess of 20%." From this we estimate "some 15 million," which is conservative.

220   carriers were able to restore functionality: Tim Large, "Cell Phones and Radios Help Save Lives After Haiti Earthquake," Reuters, January 25, 2010, http://www.reuters .com/article/2010/01/25/us-haiti-telecoms-idUSTRE60O07M20100125.

220   Digicel and Voilà, reported that they were able to operate: Suzanne Choney, "Firms Scramble to Repair Haiti Wireless Service," MSNBC, updated January 22, 2010, http://www.msnbc.msn.com/id/34977823/ns/world_news-haiti/t/firms-scramble -repair-haiti-wireless-service/#.UIBq5MVG-8B.

220   Get the towers up, get them running: Cameron R. Hume, in discussion with Jared Cohen, January 2010.

220   Donated cell towers had to be guarded: Suzanne Choney, "Firms Scramble to Repair

Haiti Wireless Service," MSNBC, updated January 22, 2010, http://www.msnbc.msn.com/id/34977823/ns/world_news-haiti/t/firms-scramble-repair-haiti-wireless-service/#.UIBq5MVG-8B.

220   Vodafone's speedy restoration of service in Egypt: "Statements—Vodafone Egypt," Vodafone, see January 29, 2011, and February 2, 2011, http://www.vodafone.com/content/index/media/press_statements/statement_on_egypt.html.

220   "We had people sleeping in the network centers": Vittorio Colao in discussion with the authors, August 2011.

222   Roshan, is also the country's biggest investor and taxpayer: "Western Union and Roshan to Introduce International Mobile Money Transfer Service in Afghanistan," Roshan, News, February 27, 2012, http://www.roshan.af/Roshan/Media_Relations/News/News_Details/12-02-27/Western_Union_and_Roshan_to_Introduce_International_Mobile_Money_Transfer_service_in_Afghanistan.aspx.

222   Roshan employs thousands: Ibid.

223   8 percent stake in *The New York Times*: Russell Adams, "Carlos Slim Boosts Stake in New York Times Again," *Wall Street Journal,* October 6, 2011, http://online.wsj.com/article/SB10001424052970203388804576615123528159748.html.

223   "I think that more than feeling just Lebanese, I feel I am part of the world altogether": Carlos Slim Helú, in discussion with the authors, September 2011.

224   growth of mobile phones in Somalia: Abdi Sheikh and Ibrahim Mohamed, "Somali Mobile Phone Firms Thrive Despite Chaos," Reuters, November 3, 2009, Africa edition, http://af.reuters.com/article/investingNews/idAFJOE5A20DB20091103; Abdinasir Mohamed and Sarah Childress, "Telecom Firms Thrive in Somalia Despite War, Shattered Economy," *Wall Street Journal,* May 11, 2010, http://online.wsj.com/article/SB10001424052748704608104575220570113266984.html.

224   functions across all three regions: "Somalia—Telecommunications Overview," Infoasaid, accessed October 18, 2012, http://infoasaid.org/guide/somalia/telecommunications-overview.

224   Only one commercial bank exists in Somalia: Mohamed Odowa, "Rebuilding Trust in Somali Commercial Banking," *Somalia Report,* May 15, 2012, http://www.somaliareport.com/index.php/post/3347/Rebuilding_Trust_in_Somali_Commercial_Banking; Dinfin Mulupi, "Opening a Bank in Somalia? Not a Crazy Idea, Says Businessman," *How We Made It in Africa* (Cape Town), June 18, 2012, http://www.howwemadeitinafrica.com/why-we-decided-to-open-a-bank-in-somalia/17530/.

225   mobile money-transfer services allow: Sahra Abdi, "Mobile Transfers Save Money and Lives in Somalia," Reuters, March 3, 2010, http://www.reuters.com/article/2010/03/03/us-somalia-mobiles-idUSTRE6222BY20100303.

225   Bahraini telecom tried to expand: Cynthia Johnston, Reuters, "U.S. Authority Tells Batelco to End Iraq Cellular Service," *Arab News* (Jeddah), July 27, 2003, http://www.arabnews.com/node/234902.

225   Somalia's mobile penetration is much higher: Author discussions with government officials in Somalia, October 2012. It is worth noting that official statistics for Somalia sometimes show a lower percentage.

226   Pirates on the Somali coast: Jama Deperani, "Somali Pirate Rules and Regulations," *Somalia Report,* October 8, 2011, http://www.somaliareport.com/index.php/post/1706.

226   In a February 2012 report, the United Nations Security Council: *Security Council*

Committee on Somalia and Eritrea Adds One Individual to List of Individuals and Entities, United Nations Security Council SC/10545, February 17, 2012, http://www .un.org/News/Press/docs/2012/sc10545.doc.htm.

226   a list of the ten functions of the state: Ashraf Ghani and Clare Lockhart, *Fixing Failed States: A Framework for Rebuilding a Fractured World* (New York: Oxford University Press, 2008), 124–166.

227   three commissions were created: *Work Package 7 on Reparations, Report of Workshop II: The Interactions between Mass Claims Processes and Cases in Domestic Courts,* Impact of International Courts on Domestic Criminal Procedures in Mass Atrocity Cases (DOMAC) and Amsterdam Center for International Law, June 18, 2010. See section by Peter van der Auweraert, presenter of "Panel Three: Iraq Reparation Schemes," pages 27–31, http://www.domac.is/media/domac/Workshop-II-report -Final.pdf.

227   A parallel authority was set up to resolve disputes: Ibid., see Discussion of the Cassation Commission, pages 28 and 30.

227   But despite their good intentions: Ibid, 29–31.

230   The Somali diaspora: France Lamy, "Mapping Towards Crisis Relief in the Horn of Africa," Google Maps, August 12, 2011, http://google-latlong.blogspot.com/2011 /08/mapping-towards-crisis-relief-in-horn.html.

231   The journalist Naomi Klein: Naomi Klein, *The Shock Doctrine: The Rise of Disaster Capitalism* (New York: Metropolitan Books/Henry Holt, 2007).

232   hundreds of thousands were killed: "Paul Farmer Examines Haiti 'After the Earthquake,'" NPR, July 12, 2011, http://www.npr.org/2011/07/12/137762573/paul-farmer -examines-haiti-after-the-earthquake.

232   The Haitian government believes: "Haiti," *New York Times,* updated August 26, 2012, http://topics.nytimes.com/top/news/international/countriesandterritories/haiti /index.html.

232   while a leaked memo: Emily Troutman, "US Report Queries Haiti Quake Death Toll, Homeless," Agence France-Presse (AFP), May 27, 2011, http://www.google .com/hostednews/afp/article/ALeqM5jELhQRaWNNs56GOlifagC5F4DSZ g?docId=CNG.699dc08a5f873f53071a317e008a7a5b.3a1.

232   "text to donate" campaign: Lindsey Ellerson, "Obama Administration Texting Program Has Raised $5 Million for Red Cross Haiti Relief," ABC News, January 14, 2010, http://abcnews.go.com/blogs/politics/2010/01/obama-administration-texting -program-has-raised-5-million-for-red-cross-haiti-relief/.

232    $43 million in aid: Elizabeth Woyke, "Yes, You Can Still Donate Money to Haiti via Your Cellphone," *Forbes,* January 12, 2011, http://www.forbes.com/sites/elizabeth woyke/2011/01/12/yes-you-can-still-donate-money-to-haiti-via-your-cellphone/.

232   Télécoms Sans Frontières: Adele Waugaman, "Telecoms Sans Frontieres' Emergency Response," presentation to the U.S. Department of State, Haiti Earthquake, July 9, 2010, United Nations Foundation and Vodafone Foundation, http://www .unfoundation.org/assets/pdf/haiti-earthquake-tsf-emergency-response-1.pdf; Tom Foremski, "Télécoms Sans Frontières—How a Simple Phone Call Helps in Haiti," *Silicon Valley Watcher,* February 4, 2010, http://www.siliconvalleywatcher.com/mt /archives/2010/02/telecoms_sans_f.php.

232   Thomson Reuters Foundation's AlertNet: Thomson Reuters, "Thomson Reuters Foundation Launches Free Information Service for Disaster-Struck Population in Haiti: Text Your Location to 4636 to Register," press release, January 17, 2010, http://

thomsonreuters.com/content/press_room/corporate/TR_Foundation_launches
_EIS.

233   delivery of funding from institutional donors: José de Córdoba, "Aid Spawns Backlash in Haiti," *Wall Street Journal,* November 12, 2010, http://online.wsj.com
/article/SB10001424052702304023804575566743115456322.html; Ingrid Arnesen, "In
Haiti, Hope Is the Last Thing Lost," *Wall Street Journal,* January 12, 2011, http://
online.wsj.com/article/SB10001424052748704515904576076031661824012.html.

233   Hundreds of thousands of Haitians: William Booth, "NGOs in Haiti Face New
Questions about Effectiveness," *Washington Post,* February 1, 2011, http://www
.washingtonpost.com/wp-dyn/content/article/2011/02/01/AR2011020102030
.html.

233   People well qualified to say what transpired: See Paul Farmer, *Haiti After the Earthquake* (New York: PublicAffairs, 2012).

233   a pattern emerging: See Jessica T. Mathews, "Power Shift," *Foreign Affairs,* January/
February 1997, http://www.foreignaffairs.com/articles/52644/jessica-t-mathews/
power-shift, about the rise of NGOs.

234   70 percent overhead in "production costs": Aly Weisman, "Invisible Children
Respond to #StopKony Viral Video Criticisms," *The Wire, Business Insider,* March 8,
2012, http://www.businessinsider.com/invisible-children-respond-to-stopkony-viral
-video-criticisms-2012-3.

234   bizarre detention: Sarah Grieco, "Invisible Children Co-founder Detained: SDPD,"
NBC 7 San Diego, March 17, 2012, http://www.nbcsandiego.com/news/local/jason
-russell-san-diego-invisible-children-kony-2012-142970255.html.

236   There already are monitoring and rating systems for NGOs: GuideStar, Charity
Navigator, GiveWell, CharityWatch, Philanthropedia, GreatNonprofits, and others
fall into the former category. Their primary goal is to facilitate better-informed giving. They range from simply aggregating relevant information such as organizations'
Form 990 tax returns (GuideStar), to engaging charities directly to collect information and analyze evidence of impact (GiveWell). While these tools are incredibly
valuable, our hunch is that they are only used by foundations and a small minority
of individuals who give in considerable amounts. A report called *Money for Good*
confirms this hunch, finding that only 35 percent of individuals do research before
giving and that those who do primarily go to the organization itself for information.
See *Money for Good II: Driving Dollars to the Highest Performing Nonprofits, Summary Report 2011,* Hope Consulting, November 2011, 9–10, http://www.guidestar
.org/ViewCmsFile.aspx?ContentID=4040.
There are already some umbrella aid organizations like InterAction, but they
have a couple of hundred members at best, which is only a fraction of the tens of
thousands of NGOs involved in some kind of relief work.

240   demonstrated this to great effect: Jason Palmer, "Social Networks and the Web Offer
a Lifeline in Haiti," BBC, January 15, 2010, http://news.bbc.co.uk/2/hi/8461240
.stm; "How Does Haiti Communicate after the Earthquake?," BBC, January 20,
2010, http://news.bbc.co.uk/2/hi/technology/8470270.stm.

240   Ushahidi volunteers in the United States: James F. Smith, "Tufts Map Steered
Action amid Chaos," *Boston Globe,* April 5, 2010, http://www.boston.com/news
/world/latinamerica/articles/2010/04/05/tufts_project_delivered_aid_to_quake
_victims/?page=1; Jessica Ramirez, "'Ushahidi' Technology Saves Lives in Haiti
and Chile," *Newsweek and Daily Beast,* March 3, 2010, http://www.thedailybeast

.com/newsweek/blogs/techtonic-shifts/2010/03/03/ushahidi-technology-saves-lives -in-haiti-and-chile.html.

240 "I'm buried": Ramirez, "'Ushahidi' Technology Saves Lives in Haiti and Chile," http://www.thedailybeast.com/newsweek/blogs/techtonic-shifts/2010/03/03/ ushahidi-technology-saves-lives-in-haiti-and-chile.html.

241 Roshan has launched a pilot program: Eltaf Najafizada and James Rupert, "Afghan Police Paid by Phone to Cut Graft in Anti-Taliban War," *Bloomberg,* April 13, 2011, http://www.bloomberg.com/news/2011-04-13/afghan-police-now-paid-by-phone-to -cut-graft-in-anti-taliban-war.html.

242 "Where people have needs": Paul Kagame in discussion with the authors, September 2011.

243 (an estimated $1 billion annual business): Aditi Malhotra, "The Illicit Trade of Small Arms," *Geopolitical Monitor* (Toronto), Backgrounder, January 19, 2011, http://www .geopoliticalmonitor.com/the-illicit-trade-of-small-arms-4273/.

244 Mali with disgruntled Tuareg fighters: Michel Moutot, Agence France-Presse (AFP), "West's Intervention in Libya Tipped Mali into Chaos: Experts," *Google News,* April 5, 2012, http://www.google.com/hostednews/afp/article/ALeqM5hJt UvEGQfSoX5Lip5M2Z7MOJIgkw?docId=CNG.90655ad2d0483083880b 2914c0ec5599.251.

245 demilitarization of tens of thousands of former fighters: *Reintegration Program: Reflections on the Reintegration of Ex-Combatants,* Multi-Country Demobilization and Reintegration Program (MDRP), September–October 2008, http://www.mdrp .org/PDFs/MDRP_DissNote5_0908.pdf.

245 "We believe that we need to put tools in the hands of ex-combatants": Paul Kagame in discussion with the authors, September 2011.

245 more than $380 million in aid: Frederick Womakuyu, "South Sudan: Nation Embarks on Disarming Ex-Combatants," *AllAfrica,* July 12, 2011, http://allafrica .com/stories/201107130081.html.

246 two hundred thousand former soldiers: Ibid.

247 In Colombia, a largely successful DDR program: Author observations while visiting the DDR program in Colombia on two separate occasions.

248 "Human-rights and justice groups can build": Nigel Snoad in discussion with the authors, March 2012.

249 Dozens of criminals: *Report of the International Criminal Court, Sixty-Sixth Session,* United Nations General Assembly, August 19, 2011, 6–7, http://www.icc -cpi.int/NR/rdonlyres/D207D618-D99D-49B6-A1FC-A1A221B43007/283906 /ICC2011AnnualReporttoUNEnglish1.pdf.

249 for many months before their trials: Susana SáCouto, Katherine Cleary et al., "Expediting Proceedings at the International Criminal Court," American University, Washington College of Law, War Crimes Research Office, International Criminal Court, Legal Analysis and Education Project, June 2011, http://www.wcl.american .edu/warcrimes/icc/documents/1106report.pdf.

249 the country's new government: *Reconciliation After Violent Conflict: A Handbook,* International Institute for Democracy and Electoral Assistance (International IDEA), 2003. See section by Peter Uvin, "The Gacaca Tribunals in Rwanda," 116–117, accessed October 19, 2012, http://www.idea.int/publications/reconciliation /upload/reconciliation_full.pdf.

250 the *gacaca* tribunal: Ibid.

Conclusion

253 "Technology": Ray Kurzweil, *The Age of Spiritual Machines: When Computers Exceed Human Intelligence* (New York: Viking, 1999), 32.

253 Every two days: M. G. Siegler, "Eric Schmidt: Every 2 Days We Create as Much Information as We Did up to 2003," *TechCrunch,* August 4, 2010.

253 only two billion people: "The World in 2010: ICT Facts and Figures," *ITU News,* December 2010, http://www.itu.int/net/itunews/issues/2010/10/04.aspx.

253 seven billion online: "U.S. & World Population Clocks," U.S. Cesus Bureau, accessed October 26, 2012, http://www.census.gov/main/www/popclock.html.

INDEX

A NOTE ABOUT THE AUTHORS

Eric Schmidt is the executive chairman of Google, which he joined in 2001. From 2001 to 2011, he served as Google's chief executive officer, overseeing the company's technical and business strategy alongside founders Sergey Brin and Larry Page. Dr. Schmidt was previously the chairman and CEO of Novell and chief technology officer at Sun Microsystems, Inc., and served on the research staff at Xerox Palo Alto Research Center (PARC), Bell Laboratories and Zilog. He is a member of the President's Council of Advisors on Science and Technology, a member of the Prime Minister's Council for Science and Technology in the United Kingdom and a life member of the Council on Foreign Relations (CFR). Elected to the National Academy of Engineering in 2006 and inducted into the American Academy of Arts and Sciences as a fellow in 2007, he also chairs the board of the New America Foundation, and since 2008 has been a trustee of the Institute for Advanced Study in Princeton, New Jersey. He lives in California.

Jared Cohen is the founder and director of Google Ideas. He is a Rhodes Scholar and the author of two books, *Children of Jihad* and *One Hundred Days of Silence,* and has written for *Foreign Affairs, Policy Review, SAIS Review, Hoover Digest, The Washington Post* and *The International Herald Tribune.* From 2006 to 2010 he served as a member of the secretary of state's Policy Planning Staff and as a close advisor to both Condoleezza Rice and Hillary Clinton, and is now an adjunct senior fellow at the Council on Foreign Relations. He is also a member of the National Counterterrorism Center's Director's Advisory Board. He lives in New York City with his wife.

A NOTE ON THE TYPE

This book was set in Adobe Garamond. Designed for the Adobe Corporation by Robert Slimbach, the fonts are based on types first cut by Claude Garamond (ca. 1480–1561). Garamond was a pupil of Geoffroy Tory and is believed to have followed the Venetian models, although he introduced a number of important differences, and it is to him that we owe the letter we now know as "old style." He gave to his letters a certain elegance and feeling of movement that won their creator an immediate reputation and the patronage of Francis I of France.

Typeset by Scribe, Philadelphia, Pennsylvania

Printed and bound by Berryville Graphics, Berryville, Virginia

Designed by Maggie Hinders